Minimills and Integrated Mills

Minimills and Integrated Mills

A Comparison of Steelmaking in the United States

William T. Hogan, S.J.
Fordham University

Lexington Books
D.C. Heath and Company/Lexington, Massachusetts/Toronto

Library of Congress Cataloging-in-Publication Data
Main entry under title:

Hogan, William Thomas, 1919-
Minimills and integrated mills.

Bibliography: p.
Includes index.
1. Steel industry and trade—United States
2. Steel minimills—United States. I. Title.
HD9515.H599 1988 338.4'7669142'0973 86-45605
ISBN 0-669-14020-1 (alk. paper)

Published simultaneously in Canada
Printed in the United States of America
Casebound International Standard Book Number: 0-669-14020-1
Library of Congress Catalog Card Number: 86-45605

The paper used in this publication meets the minimum requirements of American National Standard for Information Sciences—Permanence of Paper for Printed Library Materials, ANSI Z39.48-1984. ♾ ™

m.R. 86 87 88 89 90 8 7 6 5 4 3 2 1

Contents

Figures and Tables

Figures

Tables

Preface

The U.S. steel industry vintage 1987 is a vastly different entity than it was just five short years ago. The differences become even greater if one chronicles the massive changes that have occurred since 1977, when the industry's ability to melt steel had reached its apex. Widespread restructuring has since resulted in both a shrinkage and dispersal of steelmaking capacity. The rationalization among integrated producers contrasts with the creation, growth, and consolidation of the minimills. Recent events, however, have proven that the minimills are not an industry apart but share many of the same problems as their larger integrated counterparts. At the same time, steel technology has continued to evolve together with the needs of a steel marketplace that is being reshaped by economic change and technical advancement.

All in all, the industry has become more efficient, more highly competitive, more niche-oriented, and more capable of serving its customers with high-quality, tailor-made steels.

Much of the material used in this book's comparison of the industry's minimills and integrated mills has been gathered by means of personal interviews with the managements of 29 minimills and 14 integrated mills. Understandably, given the ongoing, substantial transformation of the industry, a number of managements are somewhat uncertain about their future plans. With changes in the industry occurring almost on a weekly basis, some of the material in this book, which was in the press in mid-1987, will doubtless be amended by the time publication occurs.

Because the book's discussion focuses on the steel industry in the United States, its tonnage designations are in *net tons* of 2,000 pounds.

Acknowledgments

The author wishes to express his appreciation and acknowledge the generous cooperation, assistance, and support received from the Association of Iron and Steel Engineers. This organization, headquartered in Pittsburgh, Pennsylvania, is an international technical society dedicated to the advancement of the technical and engineering phases of the production and processing of iron and steel.

1
Recent Steel Industry Developments

From 1982 to 1986, the steel industry was mired in a deep depression, a situation that has given rise to considerable speculation about the future of the industry. The question of whether the integrated steel companies can survive without a radical change in their structures has been widely discussed. Much has also been said about the position of the minimills in the steel industry, particularly since the management of some of these mills is interested in moving into product areas such as sheets, medium and large structural shapes, and seamless pipe, which heretofore were considered the province of the integrated companies. Some observers have even predicted that the minimill segment will expand to a point where the integrated mills will be forced to further reduce their number and size. These are interesting speculations and challenging statements which should be examined in regard to the possibility of their realization.

To pursue this examination, it is necessary to analyze a number of factors, including the decline in the steel market, the shrinkage in steel capacity, and basic changes that have taken place in both the integrated companies and minimills. These analyses are the subject of this book, which will attempt to throw some light on the future of the steel industry in terms of its structure and participants.

A number of changes in the U.S. economy have brought about significant shifts in the steel industry. To begin with, the market served by the domestic steel industry has declined significantly. In the middle to late 1970s, domestic steel shipments averaged over 90 million tons annually, with a high of 100 million tons in 1979—a figure well below the record of 111 million tons shipped in 1973.

In 1982, shipments dropped to 61 million tons, a decline of 40 percent from 1979, a very low figure for a nonstrike year in the post–World War II period. There has been a mild recovery, but shipments have not nearly reached previous high levels. They totaled 73 million tons in 1984 and 1985, but dropped to 70 million tons in 1986.

Production of raw steel from which finished products are made reached a record high of 151 million tons in 1973 and, in 1978, was 137 million tons. In 1982, it hit a low point of 73 million tons, but revived to 92 million tons in 1984, only to drop back to 81 million tons in 1986.

The capacity of the steel industry to make raw steel reached its high point of 160 million tons in 1973, a level maintained through 1977. It has since been reduced dramatically, and was 112 million tons in 1987. In terms of ironmaking, in 1974 there were 163 blast furnaces operating. During the depressed years from 1982 to 1987, the number fell to below 50. Pig iron production was 101 million tons in 1973, compared with an average of less than 50 million tons in the depressed years of the 1980s.

Although the steel industry has traditionally ranked among the least profitable manufacturing industries, it registered a profit in every year of the post–World War II period, up to and including 1981. However, since then, it has been besieged by huge losses running into the billions of dollars. These losses have been continuous since 1982 for most companies. In the long string of profitable years, there was only one year, 1977, when profits almost vanished. In that year, they were $23 million on sales of $39.5 billion, or less than one-tenth of 1 percent.[1] This performance was due to a loss that Bethlehem Steel took of $448 million in writing off plant and equipment.[2]

In 1982, losses amounted to more than $3 billion, and the hope that things would be better in the following year did not materialize. Since 1982, the industry has operated in red figures every year up to and including 1986. This has led to a gloomy outlook for the industry, particularly in view of the fact that two major companies, LTV and Wheeling-Pittsburgh, have declared bankruptcy; there are indications that others may follow. Such events were unthinkable during most of the post–World War II years.

As a consequence of the depression, integrated steel companies have declined from 20 in 1976 to 14 in 1987. The number of integrated plants also experienced a sharp drop in the same period, falling from 47 to 23. A number of plants have also been downsized as the total capacity of the industry has been brought more in line with market demand.

It should be noted that there was a slight improvement in the fortunes of the steel industry in the last quarter of 1986 and more so in the first quarter of 1987. Most of the integrated companies returned to profitability during this brief period. However, many of the problems that brought on the depression from 1982 to 1986 have not been completely solved.

Market Decline

Although the amount of steel ultimately consumed in the United States from 1982 to 1986 has remained at an annual average level of 100 million tons, much of this tonnage is imported either in the form of steel products or products made of steel. Consequently, the market available to domestic steel producers is in the range of 70 to 73 million tons, down from an annual market of 90 million tons in the middle to late 1970s. The decline has been due to imports and a sharp drop in steel purchased for automobiles, containers, oil and gas, railroads, and machinery.

Automobiles. In the 1970s the automobile industry purchased an average of more than 20 million tons of domestic steel a year, reaching a high point of 23 million in 1973. Its purchases totaled 21.5 million tons in 1977 and 21.3 million in 1978. In the depressed years of the early 1980s, shipments to the automotive industry fell as low as 9.3 million tons in 1982, when the automobile industry itself was experiencing a depression, with total motor vehicle sales falling below 7 million units, as contrasted with an all-time high of almost 13 million units in 1978.

There has been a recovery in the automotive industry. However, because of the downsizing of cars as well as the use of materials other than steel and the sourcing of parts from abroad, the demand for steel by the automotive industry did not recover. Total vehicle production rose from less than 7 million units in 1982 to 11.6 million units in 1985, a gain of 66 percent, whereas steel shipments rose from 9.3 million tons to 13 million, an increase of only 40 percent. However, the difference in shipments of steel from 1983 to 1984 is more interesting. Automobile production rose from 9.2 million units to almost 11 million units, an increase of 19.5 percent, while steel shipments to the automobile industry rose from 12.3 million tons to 12.9 million, or less than 5 percent.

The steel industry has experienced a permanent loss of approximately 10 million tons between the banner auto years of the 1970s and the 1980s, and there seems to be no reversal in sight, as the automobile industry now requires approximately 1.1 tons of steel per passenger car, as opposed to 2 tons in the 1970s, when the cars were much larger.

Containers. The container industry also reduced its steel requirements as the beer-can and soft-drink–can markets were taken over by aluminum. In 1974, the all-time record year, steel shipments to the container industry amounted to 8.2 million tons. In the subsequent years of the 1970s, shipments were over 6.5 million tons. During the 1980s shipments have trended downward from 4.5 million tons in 1982 to approximately 4 million tons in 1986. The major loss here has been in tinplate, where shipments dropped from an average of 4.5 million tons in the late 1970s to 2.6 million tons in 1986, a loss of almost 2 million tons. During the same period, tin-free steel fell from an average of 1 million tons to 935,000 tons, while black plate fell from more than 600,000 tons to 216,000 tons. This loss, particularly in tinplate, is a permanent one, for it is doubtful that tinplate will reclaim the markets lost to aluminum.

Oil and Gas. The oil and gas industry reached its high point in terms of steel demand in 1981, as it took a total of 6.2 million tons, of which oil-country tubular goods constituted 4.2 million tons. In subsequent years, when the oil boom was shattered, shipments of oil-country tubular goods fell to 700,000 tons in 1983 and, although they recovered to 1.3 million tons in 1985, there was a severe collapse to 483,000 tons in 1986. The number of rigs in operation drilling wells declined from a high of 5,430 in December 1981 to fewer than 800 in 1987.

These drops were the result of the sharp decline in oil prices; however, it is expected that there will be more rigs operating as the price of oil rises. Nevertheless, it is extremely doubtful that they will return to the record figure established in 1981. It will be fortunate if the oil-country tubular goods deliveries by domestic producers reach 2 million tons by 1990.

Railroads. Railroads have been another disappointment for the steel industry. In the late 1970s, shipments to the rail industry averaged about 3.5 million tons. In the 1982–85 period, they averaged 1.1 million tons. The drop was not only in rails, whose average went from 1.5 million tons in the late 1970s to 700,000 tons, but also in steel for freight cars, passenger cars, and locomotives, which consumed an average of 2 million tons in the late 1970s, falling to an average of slightly more than 300,000 tons in the 1982–85 period. Freight car construction hit an all-time low of 5,772 cars in 1983, down from a range of 52,000 to 95,000 in the late 1970s.

Machinery. Another industry that cut its steel requirements was machinery. It averaged about 5.5 million tons in the late 1970s, but, in the 1982–85 period, fell to approximately 2.5 million tons. This is not expected to be a permanent loss and requirements should recover to some extent as more funds are allocated to capital equipment by the major industries in the late 1980s. However, it is questionable whether the sector will return to the 5.5- to 6-million-ton figure posted from 1977 through 1979. One of the principal reasons for this uncertainty is imports of various types of machinery, which contained some 4.7 million tons of steel in 1985.

The steel industry responded to this market decline by cutting raw-steel capacity from 160 million tons to 112 million tons. Its work force dropped from 450,000 employees in 1979 to 186,000 employees in 1986.

Imports

Direct Imports. In addition to the shrinkage in steel markets, direct steel imports in the 1980s are taking a much larger share of a smaller market. In 1979, they were 17.5 million tons out of a total market of 115 million tons, or approximately 15 percent. In 1984, they were 26.2 million tons of a total market of 99 million tons, or 26 percent. In 1985, they were 24.3 million tons of a total market of 96.4

million tons, or 25 percent. In 1986, they were 20.5 million tons of a total market of 91 million tons, or 22.5 percent. The implementation of the Voluntary Restraint Agreements (VRAs) established in late 1984 reduced the import tonnage in 1986 both in terms of actual volume and percentage of market. However, both tonnage and percentage of market are much greater than they were in the 1970s.

Indirect Imports. In addition to the steel products that have come into the country, a sizeable tonnage of steel has entered in the form of products made of steel, such as automobiles and machinery. These are classified as indirect steel imports, and have a definite effect on the industry here, for if the automobiles were not imported, presumably they would have been made in the United States, and a large tonnage of steel would have been sold to their makers by the domestic producers. The same is true of machinery and other products containing steel. Automobiles and machinery are obvious and can be accurately calculated in terms of the steel that was displaced.

For 1985, an estimate has been made by the American Iron and Steel Institute that some 16.1 million tons of steel came into the country in the form of manufactured products. The tonnage of steel in exported products amounted to 8.7 million, thus leaving an imbalance of imported steel in products of 7.4 million tons.[3] This should be added to the import tonnage, since it represents lost market tonnage for the domestic steel industry.

Pricing

Part of the steel companies' problems came from their pricing practices, which during 1985 could be described as suicidal. Between the fourth quarter of 1984 and the fourth quarter of 1985, steel prices fell in some instances as much as 30 percent. These were transaction prices, which were not published and which rendered the list prices meaningless. The price reductions were a result of attempts by various companies to either gain or maintain market share.

During the same period, the steel companies instituted programs to cut costs that were quite successful. However, the reductions in costs were neutralized by the price cuts. One company stated that if the prices in the fourth quarter of 1984 were maintained through the fourth quarter of 1985, profits would have been $40 million a month. Instead, with price cuts, the company remained in an unprofitable position.

All of these factors—the shrinkage in the market, the penetration of imports, and the resultant price policy—had strong impacts on the entire steel industry and particularly on the integrated steel companies. The degree of damage, however, varied from company to company.

Steelmaking Processes

Another basic shift in the steel industry is the fundamental change in steelmaking processes during the postwar period. Until 1960, the open hearth was the dominant process, accounting for as much as 87 to 90 percent of the raw steel produced. After 1960, it began to decline; between 1965 and 1970, open-hearth output fell sharply to 36.5 percent of the total, while the basic-oxygen process maintained its dominance with around 59 percent after 1970. During this same period, the electric furnace achieved a strong position among the steel processes. In 1970, it accounted for 15.3 percent or some 20 million tons. By 1986, its contribution was over 36 percent. Table 1–1 indicates the development in the various steelmaking processes in the post–World War II period.

Much publicity was given to the rise of the basic-oxygen process, as it moved from 17 percent of production in 1965 to an average of 60 percent between 1975 and 1986. Less attention was paid to the dramatic increase in electric-furnace capacity and production over the same period. In 1975, it was almost 20 percent and, in 1986, over 36 percent. This had a number of repercussions. First, it made it possible to enter the steel business with a relatively low capital investment, since the electric furnace did not demand expensive accompanying equipment, such as blast furnaces and coke ovens. Anyone installing an electric furnace had a variety of sizes available, ranging from a capacity of 100,000 to 400,000 or 500,000 tons. Second, it operates on scrap as a "raw" material and so places a heavy demand on scrap, which was in tight supply in the boom of the mid-1970s. However, when total raw-steel output

Table 1–1
Process Mix of U.S. Raw-Steel Output, 1945–86
(percentage of total output)

	Basic Oxygen	*Electric*	*Open Hearth*	*Bessemer*[a]	*Million Net Tons of Raw Steel*
1945	———	4.4%	90.2%	5.4%	79.7
1950	———	6.2	89.2	4.6	96.8
1955	0.3%	6.9	90.0	2.8	117.0
1960	3.4	8.4	87.0	1.2	99.3
1965	17.4	10.5	71.7	0.4	131.5
1970	48.2	15.3	36.5	———	131.5
1971	53.1	17.4	29.5	———	120.4
1972	56.0	17.8	26.2	———	133.2
1973	55.2	18.4	26.4	———	150.8
1974	56.0	19.7	24.3	———	145.7
1975	61.6	19.4	19.0	———	116.6
1976	62.5	19.2	18.3	———	128.0
1977	61.8	22.2	16.0	———	125.3
1978	60.9	23.5	15.6	———	137.0
1979	61.1	24.9	14.0	———	136.3
1980	60.4	27.9	11.7	———	111.8
1981	60.6	28.3	11.1	———	120.8
1982	60.7	31.1	8.2	———	74.6
1983	61.5	31.5	7.0	———	84.6
1984	57.1	33.9	9.0	———	92.5
1985	58.8	33.9	7.3	———	88.3
1986	59.5	36.4	4.1	———	80.5

Source: American Iron and Steel Institute, *Annual Statistical Reports* for various years.
[a]U.S. Bessemer ouput was suspended in 1968.

dropped in the 1980s, the scrap supply was more than adequate. The problem was the quality of the scrap needed to make more sophisticated steel products. The electric furnace also brought the minimill into prominence as a significant part of the steel industry.

The rise of the minimill was spread over a 20-year period, from 1965 to 1985. During the latter part of that period, from 1975 to 1985, a number of minimills were built, while there was a significant decline in the number of integrated steel mills—from 47 to 23. Considerable additions were made to minimill capacity by such companies as Florida Steel, Nucor, North Star, and Atlantic, while new companies were established, including Raritan River, Chaparral, Bayou, and Auburn. Capacity was also increased, in a number of instances, by the replacement of small electric furnaces with larger ones.

Basic Definitions

There are fundamental differences between the minimills and the integrated plants. However, the original concept of the minimill has changed considerably since the early 1960s, when it was first developed. The integrated mill has also undergone some basic changes. A comparison between these two sectors of the steel industry requires a definition of each in order to make a proper evaluation.

The Minimill

The minimill, as originally conceived, had several basic characteristics:

1. In terms of size, it was generally considered to have 100,000 tons or less of raw-steel capacity.
2. Its equipment consisted of an electric furnace, wholly dependent on scrap; a breakdown mill to reduce small ingots to billet size or a continuous caster which casts billets directly from molten steel; and a bar mill.
3. The product line was usually restricted to concrete reinforcing bars, merchant bars, and, in some instances, light structural shapes, such as small angles and channels.
4. The mills are scattered over most of the country and the markets they served were usually within a 200- to 300-mile radius of the mill.

The operating procedure for minimills is relatively simple and is shown in figure 1–1. The raw material, so to speak, is scrap. This is unloaded in the scrap yards and then charged into the electric furnace. When the scrap has been melted and refined into steel, it is poured into a ladle, the contents of which are then discharged into a continuous caster. Presently, almost all of the minimills cast billets. These are then put into a reheat furnace and, when they have reached a rolling temperature, are converted into final products on the bar mill. The final product is finished hot, allowed to cool, and then prepared for shipment. Thus, the process of producing steel and

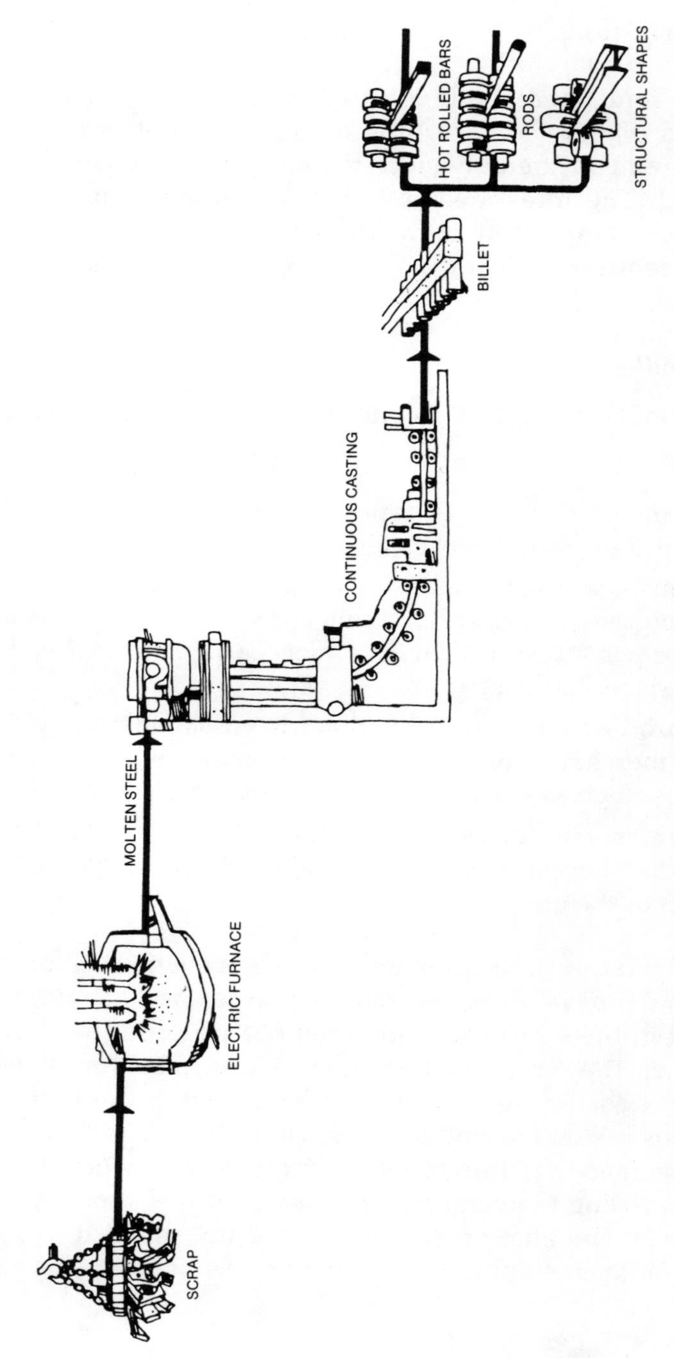

Figure 1–1. Minimill Flow Sheet: Major Steelmaking Facilities and Processing Operations

rolling it into a finished product is relatively straightforward and incorporates a limited number of operations geared to produce relatively few products. The accompanying flow sheet indicates the stages of production employed by the minimills. It should be noted that the structural shapes are usually small angles and channels with the exception of one mill, Chaparral, which produces wide-flange beams. All of the minimills finish their operation with a hot-rolled product.

A number of minimills have changed drastically from the original concept, although some have remained basically true to it. The size of the plants in terms of raw-steel production has grown significantly in almost every case. Some of the original minimills that had less than 100,000 tons of capacity in 1960 are now producing over 200,000 tons. For example, in 1960 Florida Steel Corporation's Tampa plant had one furnace with a capacity of 51,000 tons. The plant today has a rated capacity of 280,000 tons. In 1960, the Roanoke Electric Steel Corporation of Roanoke, Virginia, had a capacity of 25,000 tons. In 1987, annual capacity was approximately 500,000 tons. Other plants such as the Georgetown Steel Corporation plant went into operation in 1969 with 300,000 tons and, in 1987, had 700,000 tons. When founded in 1975, the Chaparral plant at Midlothian, Texas, had 400,000 tons; in 1987, as a result of adding a second furnace, it had a capacity of 1.5 million tons. This growth in raw-steel capacity and production is general among most of the minimills, and in that respect, they differ from the original concept.

In terms of technology, most minimills operate essentially the same type of equipment that they did at their inception. However, many electric furnaces have been replaced by larger units and have been improved through the addition of sidewall cooling panels and, in a number of cases, larger electric transformers. Virtually all of the mills now have continuous casters, as contrasted with less than half in 1968. Some of them have updated earlier versions. In terms of bar mills, improvements have been made through modifications of old units and the installation of some new ones. Additional stands have been added; new and improved cooling beds as well as new heating furnaces have been installed.

There are several mills that produce wire rods. These are placed in the minimill category, although the product is not common among minimills. The rod producers are Georgetown Steel Corporation of South Carolina, North Star Steel of Texas, Raritan River

Steel Company of New Jersey, and Keystone Steel and Wire Company of Peoria, Illinois. Atlantic Steel has a limited production, as has Florida Steel at its Jacksonville plant. Both Raritan River and North Star of Texas have rod mills that are the equal of any in the world. Georgetown has upgraded its mill so that it is almost in the same class. Therefore, it can be said that they are in the forefront of technology. They have had the field for the past few years, since the integrated mills have, for the most part, dropped out of the rod business. Other companies making a limited amount of rods include Charter Electric and Tamco.

Three minimill companies have stated their intention to produce products that were formerly the exclusive province of the integrated mills. Nucor is in the process of constructing a mill for the production of flat-rolled products, as well as a mill for the production of wide-flange beams. North Star is completing a seamless-pipe mill at its Youngstown facility. Chaparral produces wide-flange beams of up to 12 inches and is gearing up to produce larger sizes. Further, it must be noted that most minimills with improved bar mills are producing small quantities of special-quality bars.

As a result of their proposed entrance into other fields, as well as the increase in size, most of the minimills have found it necessary and desirable to expand their market area and move to a larger radius than the traditional 200–300 miles. The rod producers cover half the country.

Many minimills have branched out to the point where they become competitive with each other, much more so than with the integrated plants. Chaparral, for example, markets its medium structural beams in 40 states.

As a result of all of these changes, the minimill concept is somewhat blurred as the plants grow larger and move into different product areas. The tendency to concentration, whereby one company owns several mills, is also a radical change from the original concept where there was usually one mill to a company.

The Integrated Mill

The integrated mill is a large complex of facilities and differs from the minimill in many ways:

1. It starts with basic raw materials, such as coal and iron ore, which have to be processed before they can be charged into the

blast furnace. Coal is converted to coke and raw iron ore is upgraded by transforming it into pellets or sinter.

2. In terms of size, it is huge by comparison with the minimill. Some integrated facilities produce as much as 6 to 7 million tons of raw steel. However, others produce as little as 1 to 2 million tons.
3. The facilities include coke ovens, sinter plants, blast furnaces, basic-oxygen steelmaking converters, electric furnaces, continuous casting, and a variety of rolling mills. Some of these, such as modern hot-strip mills, are capable of producing up to 5 million tons of hot-rolled coils in a year. Other finishing facilities include plate mills, pipe mills, structural mills, galvanizing lines, and electrolytic tinning lines.
4. The quality of the integrated-mill product differs from that of the minimill. Chemical and physical specifications are more demanding. This is particularly true of sheet products that are to be formed into various parts of automobiles under the pressure of huge presses. The surface quality has become increasingly critical as has uniformity of width and gauge. These requirements have put considerable pressure on the mills to constantly upgrade their production processes.
5. The market covered by the integrated mill, in many instances, includes more than half of the United States.
6. In terms of location, the integrated mills tend to be concentrated in a few areas, principally due to raw materials. The principal area is the Great Lakes, although there are other mills located on the East Coast and in the Midwest and South.

The iron- and steelmaking processes in the integrated mill involve large tonnages and, consequently, large facilities. The prepared ore and coke are charged along with limestone into a blast furnace which, depending on its size, can produce from 3,000 to 10,000 tons of iron per day. The iron is then processed into steel in a basic-oxygen converter, after which the steel is either continuously cast into semifinished shapes (such as slabs, billets, and blooms) or poured into ingot molds where it is allowed to solidify. The ingot must be reduced in size to a useable semifinished product such as the cast slab, billet, or bloom.

The semifinished steel is further processed into a wide variety of products, including sheets, plates, pipe, heavy structural shapes, bars, tinplate, and galvanized steel. The production of these items from the semifinished state to the final product requires a number of steps. These differ markedly from the steps required by the minimill. The minimill is limited to hot-rolled products on a bar mill which are cooled and shipped. Further, the minimill does not have any iron-ore preparation, coke ovens, or blast furnaces. Thus, the integrated mill has a number of steps that must be taken before reaching the steel furnace and many more after hot rolling.

In the production of sheets, the integrated mill's major product—the semifinished steel in slab form—is reduced to a hot-rolled sheet on a hot-strip mill, which is a huge complex of equipment requiring an investment of $250 to $300 million. The hot-rolled band is then put through an acid tank, known as a pickling line, which cleanses the surface. It is then further reduced in thickness at room temperature by a cold-reduction mill, a massive piece of equipment requiring an investment of at least $150 million. Some of this steel is sold as cold-reduced sheets, while some is plated with tin to form tinplate, and some is covered with zinc to form galvanized steel. Other products made by the integrated mill require many operations as they move from the semifinished state to the finished form.

Figure 1-2 indicates the complexity and size of the integrated mill. It contrasts with the much smaller operation of a minimill in terms of raw materials processes and products produced.

In the latter part of the post-World War II period, integrated mills have undergone some basic changes. First, in the steel process, the open-hearth furnace has virtually been replaced by the basic-oxygen furnace (BOF). Continuous casting has been installed in almost every large integrated mill in the United States, so that the ingot process has been reduced to a very small amount of output. Electric furnaces have been installed in some of the plants to supplement the basic-oxygen steel process. This has been done at the Inland Steel plant; the LTV South Chicago and Cleveland plants; the Bethlehem Steel plant at Bethlehem, Pennsylvania; United States Steel's plants at Fairless Hills, Pennsylvania, and South Chicago; and the National Steel plant at Ecorse, Michigan, near Detroit. Further, some of the large integrated companies have built independent electric-furnace plants, such as the Baytown, Texas, plant of United States Steel, the Johnstown works of Bethlehem, and the Pittsburgh works of LTV Steel.

The number of integrated plants operating blast furnaces, which feed molten iron into the BOFs and into the few remaining open-hearth furnaces, has diminished considerably. Bethlehem at one time operated six integrated plants, but by 1986, this number had been reduced to three. United States Steel operated ten integrated plants, which number has been reduced to five. LTV Steel (formed from the combination of Jones & Laughlin, Youngstown Sheet & Tube, and Republic Steel) operated ten integrated plants at one time; presently, there are four functioning. Inland is the only major company that confines its operation to one integrated plant, although this will change with the new joint venture on a cold-reduction mill with Nippon Steel.

Among the changes noticeable in the integrated plants has been the trend to simplification by producing fewer products at one location. However, there are some mills (such as the Inland Steel plant at Indiana Harbor, which is Inland's only mill) that, of necessity, must produce all of the products that Inland makes. The trend toward simplification is manifest in the change in United States Steel, where the bar mill products cease to be produced at the Gary plant and have been transferred to the bar mills at Lorain, Ohio. Thus, Gary now produces only plates and sheets.

Integrated mills in a number of instances have been reduced in size. For example, the National Steel mill at Ecorse, Michigan, has been reduced to less than 4 million tons of steelmaking capacity from its previous high of 6 million tons.

Beyond minimills and integrated mills, there are other steel companies using electric furnaces that are in neither category. These are producers of specialty steel, usually stainless, alloy, or high carbon, most of which products are outside of the range of the minimill. Quanex, Carpenter, and Cyclops are examples. There are also electric-furnace plants producing large tonnages, often in excess of 1 million tons. Northwestern Steel and Wire is such a company. It has the largest electric furnaces in the world and a capacity to produce more than 2 million tons. Another company, Lukens Steel, has one of the largest plate mills in the United States and confines its production principally to plates, so that it never has been considered a minimill.

One profound effect of the development of the electric furnace has been to reduce the barrier of entry into the steel business. The

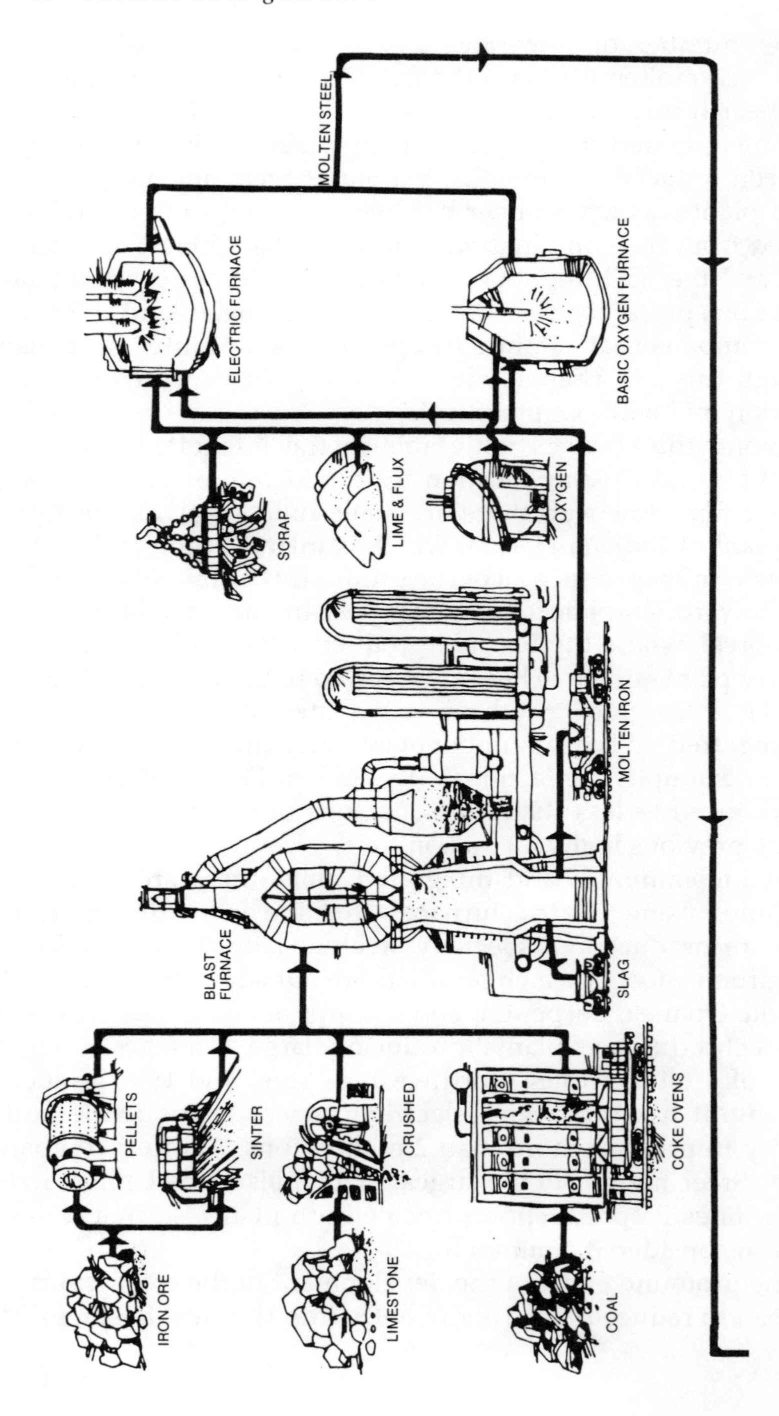
IRON ORE
LIMESTONE
COAL
PELLETS
SINTER
CRUSHED
COKE OVENS
BLAST FURNACE
SLAG
MOLTEN IRON
SCRAP
LIME & FLUX
OXYGEN
ELECTRIC FURNACE
BASIC OXYGEN FURNACE
MOLTEN STEEL

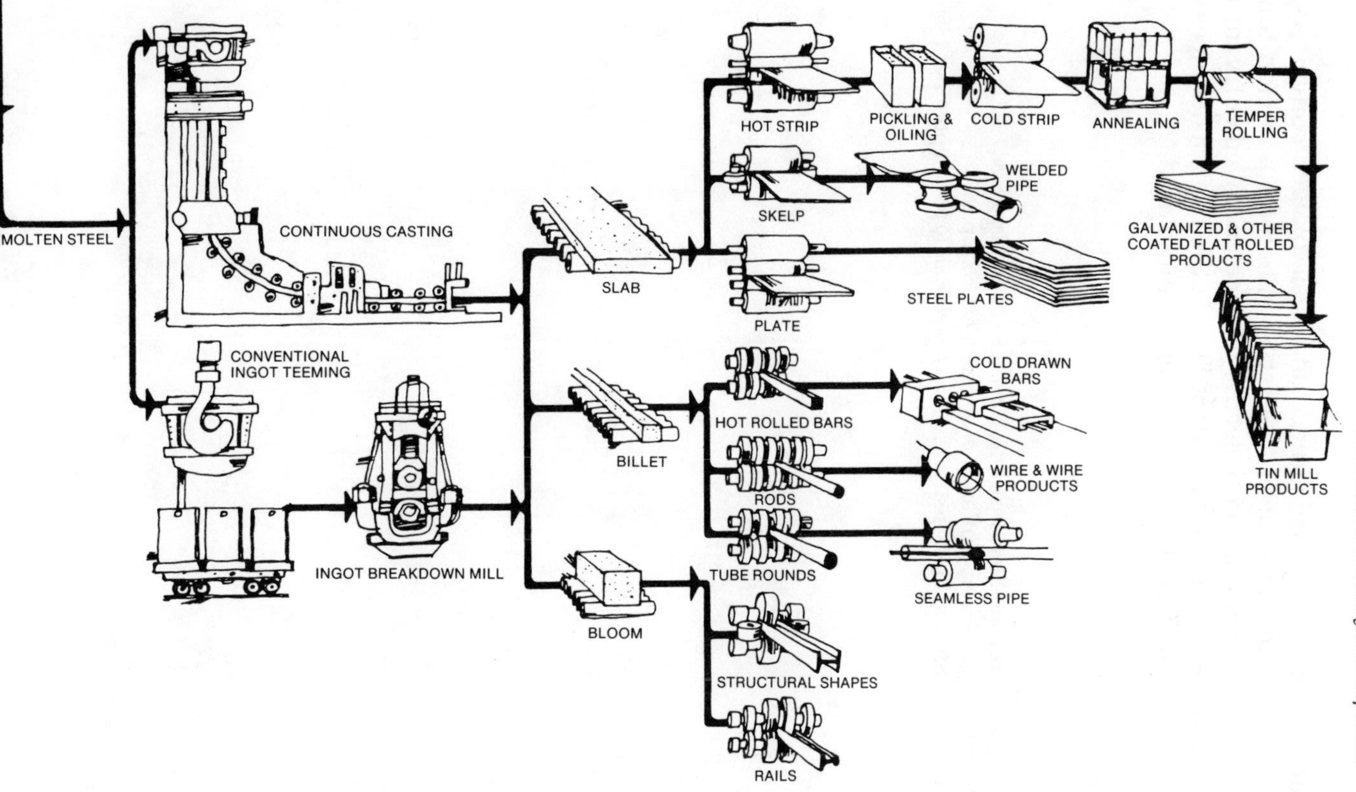

Figure 1–2. Integrated Plant Flow Sheet: Major Iron and Steelmaking Facilities and Processing Operations

integrated plant, with blast furnace and basic-oxygen converters, requires a large investment of at least $1 billion for a million-ton plant, whereas a minimill, with electric furnaces, allowed entry into the steel business for as little as $10 million in the 1960s. The price is now about $55 million, if new equipment is purchased.

Notes

1. American Iron and Steel Institute, *Annual Statistical Report for 1977,* p. 13.
2. Bethlehem Steel Corporation, *Annual Report for 1977,* p. 2.
3. American Iron and Steel Institute, *Indirect Steel Trade Report,* September 1986, p. 1.

2
Integrated Steel Companies

From 1977 to 1987, the number of integrated steel plants in the United States has fallen from 47 to 23. This decline can be attributed to a number of causes. Some were internal to the companies, while others were external and more difficult to control.

Basically, the reduced number of plants and tonnage was due principally to the market shrinkage and import penetration. As indicated in chapter 1, the consuming industries with very few exceptions took less steel, with the greatest losses registered in the automotive, container, railroad, oil and gas, and machinery industries. The integrated companies have been impacted by these factors with a degree of damage that varies from company to company and has a definite bearing on the future of each. Thus, it is necessary to examine briefly each company in light of these events.

In 1975, there were 20 integrated steel companies operating 47 plants with blast furnaces, steelmaking, and steel-finishing facilities. They were:

Alan Wood
Armco
Bethlehem
CF&I
Crucible
Cyclops
Ford
Inland
Interlake
J&L
Kaiser
Lone Star
McLouth
National
Republic
Sharon
United States Steel
Wheeling-Pittsburgh
Wisconsin
Youngstown Sheet & Tube

Between 1975 and 1985, the number of companies declined to 14 with 23 plants, as a result of closures, bankruptcies, mergers, and spin-offs. Both Alan Wood and Wisconsin were closed for financial reasons. Kaiser, under pressure to make large investments concerning pollution, shut down its blast furnaces and coke ovens and sold the remainder of its equipment. Crucible shut its blast furnaces and sold the plant to J&L. Youngstown Sheet & Tube was absorbed by J&L; subsequently, J&L and Republic merged to form LTV Steel. CF&I shut its blast furnaces and coke ovens as well as its oxygen steelmaking capacity; thus, it ceased to be an integrated mill. Cyclops closed its integrated plant at Portsmouth, Ohio. Weirton Steel was spun off by National and became an independent company, as did Gulf States, which was formerly the Gadsden plant of Republic Steel.

Some pertinent facts on these companies will help to explain their present position.

Alan Wood

Alan Wood Steel Company was one of the oldest steel companies in the United States, having been founded in 1826. In 1977, it had a capacity to produce 1 million tons of raw steel which were rolled into plate and, up until 1976, narrow strip.

In 1968, basic-oxygen steelmaking replaced the open hearth and, thus, modernized the steelmaking facilities. In subsequent years, particularly in the mid-1970s, the company's board of directors considered replacing the BOF with electric furnaces. However, the proposal was rejected when projections on the future price of scrap indicated it would be far more profitable to remain a blast-furnace/basic-oxygen iron and steel producer.

The company was successful in the post–World War II period, showing record earnings of $8.3 million in 1974. Unfortunately, it sustained a 25-day strike that year at the height of the steel boom and lost the opportunity to sell considerable tonnages of steel at high prices. The strike was costly for Alan Wood and, unfortunately, achieved little. The funds that were lost would have strengthened the financial position of the company.

In 1975 and 1976, the company sustained substantial losses of $9.4 million and $15.0 million, respectively. These were due, in large part, to the softening demand for steel plates and high operating costs, including labor and materials, particularly oil. The weakening of the market was due to a great extent to imports, which reached an all-time high, up to that point, of 19.3 million tons (over 2 million tons of which were plates, which adversely affected prices and Alan Wood).

The price deterioration and the loss of tonnage put a severe strain on Alan Wood's financial position and forced it to close its doors in 1977, thus eliminating 1 million tons of integrated steel capacity.

Armco

Armco, Inc., previously known as Armco Steel Corporation, had a raw-steel capacity in 1975 of 9.5 million tons. By 1985, this had been reduced to 6 million tons. A number of factors brought about this reduction; however, three stand out.

The first was the closure of the 1-million-ton Houston plant, which was an integrated blast-furnace and electric-arc-furnace operation, making it one of very few plants that charged hot metal into the electric furnace. In the early 1970s, the plant's open-hearth furnaces, with a 1-million-ton capacity, were replaced by four electric furnaces which were added to two existing units. The new furnaces were large, with 175-ton heat capacity, and thus, the 1-million-ton potential of the plant was increased to 2 million. Finishing facilities included a plate mill, a bar mill, and a wide-flange-beam structural mill.

In January 1984, after 42 years of operation, the Houston mill was shut down. This was inevitable, since plate imports in the Houston area reduced the mill's operation in its last year to less than 20 percent of capacity. Further, a long-term contract for energy had run out, and there was to be a substantial price increase. In commenting on the closure, Armco in its Annual Report for 1983 stated: "We saw no hope of Houston returning to profitability at any time soon."[1]

In 1987, Northwestern Steel and Wire Corporation of Sterling, Illinois, a producer of wide-flange beams, has been negotiating with Armco to purchase the structural mill at Houston.

In 1981, Armco disposed of two small electric-furnace plants, one at Marion, Ohio, and the other at Sand Springs, Oklahoma.

The third factor in the reduction of capacity was the shutdown in 1985 of the open-hearth shop at the Middletown, Ohio, plant. As a result of these closures, the company's capacity is now approximately 6 million tons.

In terms of future plans, Armco will coordinate the operation of its Ashland, Kentucky, and Middletown, Ohio, integrated plants. The steel made at Ashland will be shipped to Middletown to be rolled on the Middletown hot-strip mill, which at 86 inches in width is the widest in the United States. Some of the hot-rolled coils will then be shipped back to Ashland to be cold-reduced. This reorganization will eliminate the hot-strip mill at Ashland and reduce the plant to a one blast-furnace operation.

In addition to the integrated plants, Armco operates electric-furnace plants at Kansas City, Kansas; Butler, Pennsylvania; and Baltimore, Maryland.

In 1986, some facilities, including the seamless-pipe mill in western Pennsylvania, were written off, forcing Armco to register a heavy loss. Currently, the company has been trimmed down to its most efficient facilities. With this reduction, it operated at a higher level of capacity and returned to profitability in early 1987.

Bethlehem

Bethlehem Steel Corporation, at its peak in 1973 and 1974, had a capacity to produce 25–26 million tons of raw steel. Currently as a result of the closure of several of its plants and the reduction in size of others, the capacity stands at 16 million tons.

The major reduction in steelmaking capacity came in 1977, when both the Lackawanna plant, near Buffalo, New York, and the Johnstown plant in Pennsylvania were substantially reduced in size, resulting in a $750-million write-off and a net loss of $448 million. Subsequently, the Lackawanna plant was closed down, except for the coke ovens, the bar mill, and a galvanizing line. The Johnstown plant was reduced in tonnage significantly, as two electric furnaces replaced the blast-furnace/open-hearth complex. Currently, the capacity at Johnstown is approximately 1.3 million tons, a reduction from 2.4 million tons. Bethlehem also disposed of two electric-furnace plants on the West Coast, closing one in Los Angeles and selling another in Seattle.

As of 1987, Bethlehem operates three integrated plants at Bethlehem, Pennsylvania; Sparrows Point, Maryland; and Burns Harbor, Indiana. The company plans to remain predominantly a steel producer and has recently installed continuous-casting units at Sparrows Point and Burns Harbor, so that over 80 percent of its flat-rolled products are continuously cast.

In 1985, Bethlehem officials gave serious thought to abandoning the iron and steelmaking sector of the plant at Bethlehem. The plan was to make slabs at Sparrows Point and feed them into the finishing facilities at Bethlehem. However, in labor negotiations in early 1986, the union agreed to a reduction of 550 jobs at the Bethlehem plant. Under these circumstances, the plant could operate at a profit, and it has remained intact as an integrated steelmaking unit.

As indicated, Bethlehem has completed a major capital expenditure program with the installation of two continuous-casting units at a total cost of $540 million, which was financed off–balanced-sheet. The one major expenditure that Bethlehem faces is the renovation of the hot-strip mill at Sparrows Point, which will have to be undertaken for the company to be competitive at that plant. The cost will be approximately $150 million.

The Burns Harbor plant is the latest integrated mill built in the United States. It is principally a producer of flat products, including plates and a variety of sheet products, such as hot-rolled, cold-rolled, and galvanized sheets. The recent addition of a continuous caster permits the plant to continuously cast virtually all of its sheet product. In terms of technology and competitive position, it has a very high rating.

The Sparrows Point plant of Bethlehem was at one time considered to be one of the largest, if not the largest, plant in the world, with a capacity of over 8 million tons. In addition to its markets in the East, it supplied steel to the West Coast by water via the Panama Canal. Subsequent competition from imports reduced the West Coast market substantially. Additional factors contributed to the decline of Bethlehem's market in the East, and thus the Sparrows Point plant capacity was significantly reduced. Recent improvements at the plant include a continuous caster and the construction of a blast furnace capable of producing 10,000 tons a day. This new furnace, and Inland's no. 7 furnace, are the two largest furnaces in North America.

CF&I

In 1957, CF&I Steel operated five plants, two of which (the flagship plant at Pueblo, Colorado, and the Wickwire Spencer plant near Buffalo, New York) were integrated. The other three plants were not integrated. One at Claymont, Delaware, operated open-hearth furnaces with a cold-metal charge, consisting principally of scrap. Another at Birdsboro, Pennsylvania, had a blast furnace and a foundry; a third, the Roebling plant at Trenton, New Jersey, was originally a cold-metal, open-hearth shop. In the mid-1960s, these Roebling furnaces were replaced with three electric furnaces.

During the late 1950s and early 1960s, CF&I seriously considered integrating the Claymont plant by installing coke ovens, a blast furnace, and a basic-oxygen shop. Several million dollars were invested in a breakwater at the plant, but after some years, the project was abandoned.

Since the mid-1970s, CF&I has operated only the works at Pueblo, Colorado. All others have been closed or sold. In 1983, the Pueblo plant consisted of coke ovens, four blast furnaces, basic-oxygen converters, two electric furnaces, and finishing facilities. The principal products were oil-country tubular goods, rails, and wire. The demand for oil-country tubular goods and rails has, as indicated, declined severely. This was unfortunate since the rail mill was rebuilt in the late 1970s to allow for the production of longer rails. Rail demand fell off as a consequence of shrinkage in rail track in the United States, from 311,000 miles in 1975 to 242,000 miles in 1985. The drop in oil-country tubular goods demand resulted in the cancellation of plans to build a new seamless-pipe mill at Pueblo.

The Crane Company, which acquired CF&I in 1969, tried to sell the Pueblo plant in the early 1980s. Since it did not succeed in this attempt, the company permanently shut down the blast furnaces, coke ovens, and basic-oxygen converters in December 1983. Subsequently, in May 1985 the Crane Company spun off CF&I to Crane stockholders, and it became independent.

As of 1987, the plant operates two electric furnaces, a rail mill working at half of its capacity, and a seamless-pipe mill at much less than half of its capacity. With the closure of the basic iron and steelmaking facilities, CF&I has been removed from the list of integrated companies.

Crucible

Crucible Steel was a specialty-steel producer. However, at its Midland, Pennsylvania, plant, it operated two blast furnaces and two oxygen converters, which provided steel for sheet mills. The company was unusual insofar as almost all specialty-steel companies make their steel in electric furnaces.

The blast furnaces and oxygen steelmaking facilities were abandoned in 1981 to be replaced by two large electric furnaces. These and the rest of the steel plant were sold to J&L in 1983, but after the merger of that company with Republic to form LTV, the plant was again sold to the employees and currently operates as J&L Specialty Steel Company.

Cyclops

Cyclops negotiated with Detroit Steel Corporation in 1970, acquiring the integrated plant of that company, located at Portsmouth, Ohio. The plant consisted of coke ovens, 1 active blast furnace, 5 open hearths, a 56-inch–wide hot-strip mill, and a 3-stand cold mill. The open hearths were relatively large furnaces capable of producing 320 tons per heat. The strip mill was not a modern mill, but was judged to be adequate for Cyclops's purpose, which was to feed steel strip to its Sawhill tubular division.

After the steel boom of 1973 and 1974, the market deteriorated, so that the Portsmouth plant was not able to operate profitably and, except for the coke ovens and blast furnace, was shut down in 1977. The blast furnace was shut down in 1980, at which time the coke ovens were sold to McLouth Steel Corporation. However, when McLouth went into bankruptcy, the ovens were turned over to a trusteeship and are currently producing coke for Rouge Steel, a subsidiary of Ford Motor Company.

As of 1987, Cyclops continues to operate a steel plant at Mansfield, Ohio. It is based on electric furnaces and purchased semifinished steel.

The plant at Portsmouth, which was shut down in 1980, was at best a marginal facility. When purchased in 1970, the Cyclops management planned to make changes in the plant to bring it into a competitive position. These changes and improvements would have required an investment of some $40 million. But, with the drop in steel demand, this investment was not made, and over 1 million tons of integrated steel capacity were eliminated.

Ford (Rouge Steel)

In 1975, Ford Motor Company operated an integrated steel plant at River Rouge, which had a raw-steel capacity in excess of 3 million tons. It consisted of coke ovens, blast furnaces, and basic-oxygen converters, with two large electric furnaces under construction.

The company installed a new 68-inch-wide hot-strip mill in 1973 and 1974. However, its cold-reduction facilities were not adequate to provide the necessary quality in cold-reduced steel to meet the Ford Motor Company's requirements. As a consequence, the amount of the steel plant's production sent to the motor company dropped from 65 percent to 35 percent.

In 1982, Ford entered into negotiations with Nippon Kokan Steel Company of Japan in an effort to sell the plant. There were a number of problems, including the fact that the United Autoworkers had organized the workers, who were receiving some $5.00 per hour more than workers in steel plants organized by the United Steelworkers of America. Although the Ford organization made a commitment to renegotiate the wage level and agreed to accept 50 percent of the plant's steel production, Nippon Kokan terminated negotiations.

At this point, Ford Motor Company decided to convert the plant from a division to a subsidiary and changed the name to Rouge Steel. Following this, a substantial investment of $300 million was made to install a continuous caster and upgrade the cold-reduction mill, as well as to rebuild one of the coke-oven batteries. Part of the $300-million investment represented Rouge's portion of a joint venture with United States Steel to build a continuous-electrolytic-galvanizing line capable of producing 700,000 tons annually. By 1992, Rouge plans to spend some $400 million for a number of items, including a blast-furnace rebuild and improvements to the hot-strip mill.

Gulf States

Gulf States Steel is a newly created company, formed in 1986 with the purchase by the Brenlin Group of the Gadsden, Alabama, plant of the former Republic Steel Corporation. A condition of the Republic and J&L merger was that the Gadsden plant had to be spun off. This was no hardship since considerable thought had been given to abandoning the plant before the merger.

Before the Brenlin Group consummated its purchase, there was an effort on the part of the employees to purchase the plant and operate it as an employee stock ownership plan. This did not succeed. The Brenlin Group was able to put facilities back in operation and reach an agreement with the employees to reduce employment cost to $16 an hour.

At present, the Brenlin Group plans to continue to operate the blast furnaces and BOF in order to produce steel for plates and sheets. Whether this plant will continue as an integrated unit remains to be seen. If it does not, it will probably operate as an electric-furnace steel mill.

Gulf States facilities consist of coke ovens, one operable blast furnace, basic-oxygen steel converters, a slabbing mill, a plate mill, a hot-strip mill as well as a cold-reduction mill, and a galvanizing line. The company has succeeded in operating its facilities effectively and will make investments in plant and equipment as they are needed. At present, an investment is contemplated to install a coil box on this hot-strip mill.

Inland

Inland Steel Company is the only *major* integrated company of the 6 in the United States, that operates a single plant. All of its capital investment in steel until 1984 had been made at the Indiana Harbor plant, which currently has 6.5 million tons of steelmaking capacity. It has 9 blast furnaces, 1 of which, no. 7, is capable of producing 10,000 tons of iron per day. It and Bethlehem's "L" blast furnace at Sparrows Point represent the 2 largest furnaces on the North American continent.

In 1986, the open-hearth shop at Inland was officially abandoned and capacity reduced from 9.1 million tons to 6.5 million tons. Inland is primarily a producer of light, flat-rolled products, although it does have a significant output of wide-flanged structural beams as well as bars.

In 1987, Inland and Nippon Steel Corporation of Japan signed an agreement for a joint venture to build a completely new cold-reduction mill complex. The investment will be approximately $400 million, of which Inland is to contribute $90 million and Nippon Steel $60 million; the remaining $250 million will be provided by Japanese trading companies, including Mitsui, Mitsubishi, and Nisshio-Iwai. When this materializes, it will be Inland's second joint venture. The first was the continuous-electrolytic-galvanizing line with Bethlehem Steel and Prefinish Metals to furnish material for the automobile industry.

The new cold-reduction complex will be located at New Carlisle, Indiana. This will be the second major facility installed by Inland located outside of the Indiana Harbor plant. The Indiana Harbor works is one of the most efficient integrated steelmaking facilities in the United States. An indication of this is its man-hour-per-ton capability. In 1987, it achieved 4.0 to 4.5 man hours per ton shipped. The ultimate objective is 3 man hours.

Along with most of the steel industry, Inland suffered losses during the 1982–85 period. However, late in 1986, it regained profitability. The soundness of Inland's financial position can be attested to by the fact that its pension liabilities are fully funded.

Acme (Interlake)

Acme Steel Company is the revival of a name which disappeared when Acme Steel and Interlake Iron were merged in 1964 to form Interlake Steel. This merger brought together the Acme Steel Company, which operated basic-oxygen furnaces supplied with hot metal from a cupola, and Interlake Iron Company, which had blast furnaces in South Chicago, Toledo and Erie, Pennsylvania. The furnaces at Erie and Toledo were subsequently shut down, and the company operated on molten pig iron produced in the two South Chicago blast furnaces and converted to steel at the Acme plant by the oxygen converter. The product was narrow strip.

Up to 1986, Interlake Steel functioned as a part of a larger conglomerate. Finally, in 1986, the steel segment of the company was spun off, and the name Acme Steel was reactivated. The steel segment, with a capacity to ship 500,000 tons of products, has been relatively profitable and, from all indications, should continue to be so.

Kaiser

Kaiser Steel Corporation, located at Fontana, California, was founded in World War II to produce plates for shipbuilding on the West Coast. After the war, its plate operation was expanded by adding pipe, strip, sheets, tinplate, and galvanized sheet products. With its four blast furnaces and an open-hearth steelmaking shop, it had a capacity of over 3 million tons. In the late 1950s, it was one of the first major companies to install a basic-oxygen converter.

The company fared reasonably well through the mid-1970s. However, toward the end of the decade, a number of problems arose which ultimately resulted in the closure and sale of the facilities. In 1975, Kaiser embarked on a capital investment program which was to ultimately cost $233 million. The facilities installed included a new basic-oxygen shop, which replaced the open hearth, and a continuous slab caster. These were finished in 1978. The annual report of 1975 states:

> This program supplements and replaces some features of the facilities run-out program announced in 1974. It essentially involves the replacement of some of the exisiting open-hearth steelmaking shop with a more efficient, two-unit basic-oxygen steelmaking shop and the installation of a continuous slab caster.
>
> The new program is expected to reduce operating costs, increase the efficiency of the steelmaking operation, and avoid the necessity of investing additional capital in the old open-hearth facilities to bring them into compliance with increasingly stringent air-pollution requirements.[2]

By 1980, the corporation was slipping into a difficult financial position and held discussions with LTV Corporation of Dallas, owner of J&L Steel Corporation, on the possibility of merging. These came to naught. Faced with continuous losses from the steel operation, the Kaiser board of directors undertook a detailed study of the outlook for the company as to "whether it would be in the best interest of the company and its stockholders to continue in business or to voluntarily liquidate."[3] The opinion to liquidate was rejected, and the company continued to function. However, large expenditures to maintain compliance with the EPA pollution regulations made it more and more difficult for Kaiser to continue, and steel operations were closed down in 1983.

Other contributing factors were the uneconomical operation of the Eagle Mountain ore mine, since more overburden had to be removed to reach the high-grade ore. There were proposals to import ore, but the costs of importing and handling—as well as shipping it from the West Coast to Fontana—made the proposal untenable. Kaiser also had a higher wage cost than the rest of the industry, and a number of its facilities were over staffed.

After the plant had been closed down for approximately a year, most of the facilities were sold to a group of investors from Japan, Brazil, and the United States. The blast furnaces and coke ovens were not included in the sale and were subsequently demolished. Thus, Kaiser dropped from the ranks of integrated steel companies.

The new firm was called California Steel Company, with Wilkinson and Associates controlling 50 percent; Kawasaki Steel of Japan, 25 percent; and CVRD (the Brazilian iron ore company), 25 percent. The new company with no steelmaking capacity had to purchase slabs and, in 1985, imported approximately 500,000 tons from the Tubarao plant in Brazil. In 1986, Wilkinson sold its interest to the other two partners so that, currently, the company is owned by Kawasaki and CVRD. The output now consists of plates, sheets, and galvanized sheets.

Lone Star

Lone Star Steel came into existence in World War II with a blast furnace, which was needed to produce pig iron in Texas for the war effort. After the war, steelmaking facilities, as well as finishing facilities for pipe, were added, and Lone Star became a significant factor in the production of pipe with the emphasis on oil-country tubular goods. The principal steel facility was an open-hearth shop. However, within the past few years, small electric furnaces were added, bringing its capacity to almost 1 million tons.

The company prospered while the oil industry boomed. This was particularly true in 1981. Since that time, with the drop in demand for pipe and a very large tonnage of imports, Lone Star gradually cut back until, in 1986, the blast furnace and open-hearth facilities were shut down. This is not considered permanent; therefore, the plant is still listed as an integrated facility. However, there is some doubt as to whether the iron and steelmaking facilities will be reactivated in the near future.

At present, it is using electric furnaces and wishes to buy substantial tonnages of semifinished slabs to make flat-rolled products as well as pipe. In 1987, the company hopes to convert some 400,000 tons of slabs into hot-rolled sheets.

LTV

LTV Steel was created in 1984 as a result of a merger between J&L Steel Corporation (a subsidiary of LTV of Texas) and Republic Steel Corporation. Prior to this merger, J&L absorbed Youngstown Sheet & Tube in December 1978. Youngstown Sheet & Tube had two plants in the Youngstown area (the Campbell works and the Brier Hill works) and another plant at Indiana Harbor, Indiana, near Chicago. This last facility was acquired in the early 1920s, when the Steel and Tube Company of America was taken over by Youngstown Sheet & Tube.

The Indiana Harbor works, during the latter part of the post–World War II period, was the beneficiary of most of the capital expenditures made by Youngstown Sheet & Tube. A new 32-foot–diameter hearth-blast furnace was constructed and put in operation in 1967 with a capacity to produce 4,000 tons of iron a day, large by the standards of those times. A basic-oxygen steelmaking plant was installed in 1968 with a 3.5-million–ton capacity. In addition, an 84-inch–wide, modern hot-strip mill, capable of producing 4 million tons of hot-rolled strip, was constructed in 1968. The company also installed a 6-stand, 4-high cold-reduction mill, 1 of only 3 in the United States. These major investments were made at Indiana Harbor, since it was in the midst of the Chicago area market, considered the largest and most lucrative in the United States.

The plants at Youngstown received very little attention in the way of capital investment. It was not until the mid-1970s that the company made a decision to replace the open-hearth furnaces with basic-oxygen converters. This was never carried out, since the recession in steel after the 1974 boom hit the company very hard, and, before it could savor the recovery of 1978, it was on the verge of bankruptcy.

J&L's move to merge with Youngstown Sheet & Tube was not viewed favorably by the U.S. Department of Justice. There were indications that the department would reject the merger proposal. However, in a lengthy meeting with Attorney General Griffin Bell in June 1978, the plight of Youngstown Steel & Tube was graphically portrayed, and the attorney general decided to permit the merger. With the merger's formal conclusion in December 1978, the Indiana Harbor plant became a significant part of the J&L Steel complex.

The two plants in the Youngstown area (Campbell and Brier Hill) were almost completely abandoned by the end of 1979. All that remained was the seamless-pipe mill located at the Campbell works. Thus, two more integrated plants were dropped from the list.

The merger of J&L and Youngstown put an end to a plan to construct two modern blast furnaces in the Youngstown area, which would be financed and operated as a joint venture involving three partners, Youngstown Sheet & Tube, Republic, and Sharon. The cost of the project, which included coke ovens, was in the area of $700 million and could have provided the Youngstown area with modern ironmaking facilities. However, once Youngstown Sheet & Tube ceased to exist as an independent company, the two remaining prospective partners had no wish to carry out the project.

In 1984, after Republic Steel had written off its integrated plant at Buffalo, with 1 million tons of capacity, J&L agreed to the merger. This was at first rejected by the Department of Justice, but after considerable negotiations, it was accepted, provided the resulting entity would spin off the Gadsden, Alabama, plant of Republic as well as the cold-finishing, stainless-steel facility of Republic at Canton, Ohio. The merger was consummated in mid-1984 and resulted in the second largest steel company in the United States. Its nominal capacity to produce 23 million tons of raw steel placed it ahead of Bethlehem.

After the merger, the company fared very poorly, as did most of the other integrated steel producers. In July 1986, it sought protection under chapter 11 of the bankruptcy laws. Since that time, a number of changes have taken place with regard to the company's facilities.

In the Youngstown area, the blast furnaces, formerly operated by Republic Steel, have been abandoned, and the seamless-pipe mill, which was modernized by J&L before the merger at a cost of some $70 million, has been mothballed. The $70-million investment was designed to make the mill a world-class, competitive facility. It is interesting to note that efforts to sell the mill before it was shut down were unsuccessful. It was offered to North Star, which has a minimill in Youngstown and was constructing a seamless mill. North Star refused to buy it and went ahead with the construction of its own mill.

In the Pittsburgh area, two plants, formerly operated by J&L, have been practically eliminated. The plant at Aliquippa (which was a fully integrated operation with 5 blast furnaces, basic-oxygen steelmaking, and a number of finishing facilities) has now been virtually abandoned. All that remains is a cold-reduction mill and a tinplating operation. The bar mill, which for decades produced the well-known J&L "junior" beam, has been sold. The plant employed 10,000 people in 1981, but now employs fewer than 1,000. The Pittsburgh works also has been shut down. At one time, this was a fully integrated plant with coke ovens, blast furnaces, and open-hearth furnaces, as well as some finishing facilities. The blast furnaces and open hearths were eliminated in the late 1970s and replaced by 2 very large 350-ton electric furnaces. These have now been shut down, as have the rolling mills. All that remains of the Pittsburgh plant are coke ovens, which supply coke for the Indiana Harbor blast furnaces. The South Chicago plant's iron and steelmaking facilities were shut down, leaving its coke ovens and some finishing facilities in operation.

LTV has some further restructuring, but as it stands now, the company, which upon its founding in 1984 operated five integrated steel mills, now operates three, the plant at Warren, Ohio, the Cleveland plant, and the Indiana Harbor plant near Chicago. As a result of its bankruptcy status, LTV will possibly close more plants or parts of plants. There are two possibilities: (1) the Warren, Ohio plant and (2) the Cleveland plant (formerly two plants, one Republic and the other J&L, which are literally adjacent to one another) will most probably witness some rationalization. The capacity to produce raw steel will drop to 14 million tons.

In 1985, LTV entered into a joint venture with Sumitomo Metals of Japan to construct a continuous electrolytic galvanizing line. This was completed in 1986 and is operating successfully.

McLouth

McLouth Steel Company as an integrated operation was formed after World War II, during the steel shortage, and was financed in part by General Motors. Its raw-steel capacity was approximately 2 million tons, and it depended for a very large part of its sales on the automobile industry. The company prospered and was among the first to install an oxygen converter and continuous casting.

In 1980, principally as a result of the heavy imports and the drastic drop in automobile production from 11.5 million vehicles to 8 million, McLouth found itself in grave difficulty. With heavy losses projected, it filed a petition under chapter 11 of the Federal Bankruptcy Code in December 1981. As part of its survival strategy, McLouth was able to negotiate a contract with the union for $18 an hour employment cost, which was considerably below the going rate of other companies.

In 1982, as the company faced liquidation, Cyrus Tang made an investment which kept it intact, as he took over ownership. Since that time, with the steel price situation being particularly bad, McLouth has failed to turn its business fortunes around. Presently, it operates one blast furnace, and when this goes down for relining, it will be able to operate the second furance for possibly two more years before major investments must be made. It is possible that a decision will be made to replace the blast furnace with one or two more electric furnaces and have the company dropped from the list of integrated plants.

McLouth Steel changed its name to McLouth Steel Products in a recent reorganization. Presently, the principal owner, Cyrus Tang, wishes to sell his 65 percent share to the employees for $1 million. However, there are a number of conditions attached, such as that McLouth would continue to use Tang as their scrap broker and also sell their products to Tang's service center. Further, Tang wants an option to buy back 20 percent of the company for $5 million if it returns to profitability. The union has rejected this plan. However, there is a very strong probability that another offer will be made.

McLouth has been in financial trouble, and it is possible that its survival will be dependent on the employees purchasing it on an Employee Stock Ownership Plan (ESOP).

National

National Steel has undergone a drastic change in its corporate organization and facilities. Since 1974, the company has reduced its size by 50 percent, dropping from 12 million tons of steel capacity to 6 million. This was accomplished in two major steps.

First, in 1981, the plant at Great Lakes, which formerly had two basic-oxygen steelmaking shops and approximately 6 million tons of capacity, was reduced to about 3.5 million tons through the closure of one oxygen-steelmaking shop. The number of active blast furnaces were also reduced from four to two.

In 1984, National spun off the Weirton division, which was purchased by the employees. This had approximately 3 million tons of steel capacity. Thus, National was left with the reduced plant at Great Lakes and the plant at Granite City, Illinois, which was acquired when National merged with Granite City in 1971. The merger was permitted on the basis of the "failing company doctrine," since Granite City demonstrated that within a few years it would be in serious financial trouble.

National also operates a finishing facility, known as Midwest Steel, built in 1961 near Chicago. In 1974, plans were drawn up to make this a completely integrated plant. However, after the steel boom collapsed, the plans were abandoned.

National Steel has also undergone major changes in its corporate organization. In 1983, the diversification of National Steel's operation was recognized by the formation of a new corporate entity, National Intergroup. Prior to September 1983, National Steel was the umbrella company for aluminum, finance, energy, and steel service centers. In 1983, the company set each of these up as a profit center, standing on its own, with the largest being National Steel.

In 1984, after an unsuccessful attempt to merge with United States Steel, National Intergroup sold a half interest in its steel operations to Nippon Kokan Steel Corporation of Japan. This is the largest stake that the Japanese have in the U.S. steel industry. Furthermore, in 1986, the president of National Steel retired and his place was taken by a Japanese, Mr. Kokichi Hagiwara.

National sought to abandon the steel business in the attempted merger with United States Steel. It now owns 50 percent of the steelmaking operations and has acquired a number of nonsteel activities, which the company feels are more lucrative than the steel business in terms of return on investment.

In order to save $60 million, National Steel made a decision at its Great Lakes plant not to rebuild coke-oven battery no. 5, which was taken down in December of 1986. It will continue to operate no. 4 through 1991, when the permit expires. Consequently, it must buy some 50,000 tons of coke out of an 80,000-ton requirement. It is possible that by 1991 battery no. 5 may be rebuilt, but the decision currently is to put that off because of inadequate financing. In terms of blast furnaces, there are two operating, designated B and D. C has been abandoned, and A is being relined and will be ready for operation in the fourth quarter of 1988, at which time it will replace D, which will go down for relining. Production projects at the Great Lakes plant include a continuous electrolytic galvanizing line and a continuous caster. The electrolytic line is in operation and the caster should be in operation in 1988.

Sharon Steel Corporation

Sharon Steel Corporation is owned by NVF, a company in the Victor Posner Group. Acquired by Posner in 1968, the plant is integrated insofar as it operates blast furnaces, a basic-oxygen converter, electric furnaces, and a variety of finishing facilities. It does not have coke ovens and must purchase its entire coke requirement.

Currently, there are two blast furnaces standing but only one is in operation. Future plans call for a one blast-furnace operation. The idle blast furnace is being relined so that it can be put into operation when the lining on the current operating furnace wears out. Sharon produces a significant tonnage of alloy and high-carbon steel, along with coated material and the normal carbon sheet products.

Like a number of companies, Sharon has experienced financial problems, which have forced it to seek the protection of Chapter 11 of the bankruptcy laws.

United States Steel

United States Steel Corporation changed its corporate structure in 1986 by establishing a holding company called USX, which has United States Steel as a wholly owned subsidiary. Other subsidiaries include Marathon Oil and Texas Oil and Gas.

During the past few years, the company has undergone a number of structural changes and has reduced its raw-steel capacity significantly. In the mid-1970s, its raw-steel potential was 38 million tons. In 1987, it is approximately 19 million tons. Two large cuts were made, the first in December 1983, when some 5.2 million tons of raw-steel capacity were closed down. The second was in February 1987, when the indefinite idling of 7 million tons of capacity was announced.

The area most seriously affected was the Pittsburgh region, where the number of blast furnaces was reduced from 12 in 1975 to 3 in 1987. The Duquesne plant was closed as well as the Homestead plant, leaving the Edgar Thomson-Irvin works as the only remaining integrated plant.

In Chicago, the South Chicago plant was reduced from a fully integrated operation with over a 5-million–ton capacity to an electric-furnace facility with less than 1 million tons. The Gary plant has been simplified in terms of its product line and now produces only sheets and plates. Formerly, it produced bars and rails. Rails were dropped as a product and the bars were tranferred to the Lorain, Ohio plant.

Other facilities that were closed include the Cayahoga works in Cleveland, which produced wire and narrow strip, and the Johnstown plant, which produced castings. Both have been sold and have been put back in operation.

Prior to these announcements, the Youngstown plants, designated the Ohio works and the McDonald works, were closed down in 1979. This eliminated 4 blast furnaces and over 2 million tons of raw-steel-making capacity. Earlier in the decade (in 1971), the iron- and steelmaking and -finishing facilities at Duluth, Minnesota, were closed down. This entailed the loss of 1 million tons of integrated steelmaking capacity. The coke ovens at that plant continued to operate until 1978, when they were shut.

In the February 1987 announcement, the Geneva works was indefinitely idled with its three blast furnaces and more than 2 million tons of steelmaking capacity. In addition, most of the Baytown, Texas plant, an electric-furnace steelmaking facility, was indefinitely idled. This applied to the electric furnaces, the continuous casters, and the pipe mill. The plate mill was not included in the announcement.

In April 1987, United States Steel announced that the Geneva plant would be closed, and formal notification was sent to the union on June 1. At the same time, an announcement was made or a tentative agreement between United States Steel and a group in Utah to buy the steel mill.

The Fairless works, with approximately 3 million tons of raw-steel-making capacity, will not be permanently shut down for the duration of the labor contract (through 1991). However, the iron-and steelmaking facilities at Fairless will be in jeopardy after that time. The coke ovens at Fairless have already been closed down, and coke is being supplied from Clairton in the Pittsburgh district. The present facilities of the Fairless plant include a sinter plant, 3 blast furnaces, an open-hearth shop, a hot-strip mill, and a cold-reduction mill, as well as other finishing facilities. The iron- and steelmaking equipment is over 35 years old and in need of replacement. To rehabilitate and modernize the hot end of the plant would involve an expenditure of over $800 million. A new sinter plant would be needed as well as extensive work on the blast furnaces and a replacement of the open hearth with basic-oxygen steelmaking. Further, Fairless does not have a continuous caster and that too would have to be provided. The cost of under $1 billion would seem to be prohibitive, and thus, after the four-year contract has expired, it would seem that the iron and steelmaking part of the Fairless works would be eliminated, further reducing United States Steel's capacity by 3 million tons.

In 1983, there was an attempt to shut down the iron- and steelmaking facilities at Fairless and supply slabs by forming a joint venture with British Steel. This came to naught, since British Steel could not furnish slabs at an acceptable cost. When and if the iron- and steelmaking facilities at Fairless are closed down, either slabs or hot-rolled coils will be brought in from another location. The finishing facilities will be maintained, and United States Steel will provide its customers in the East with the same tonnage that had been sold in the mid-1980s.

The Geneva works, near Salt Lake City, Utah, was built during World War II. Its location was determined by military strategy—it was to be out of range of possible Japanese bombing raids. The plant has coke ovens, blast furnaces, and an open-hearth shop with somewhat over 2 million tons of raw-steel–making capacity. Its hot-strip mill was originally designed as a 132-inch–wide plate mill and does not have the competence of the modern hot-strip mill. The plant's primary function is to supply hot-rolled coils to the finishing facilities located at Pittsburg, California, where they are transformed into cold-rolled sheets, galvanized sheets, and tinplate.

Specifications for sheet production demanded by the customer have changed significantly since the Geneva plant and the Pittsburgh, California, finishing facilities were installed. The latter were constructed in the early 1950s. At the time they were installed, they represented the latest technology, and although the facilities are adequate for many items, they are now inadequate for a number of others. Therefore, if United States Steel is to remain competitive in the West Coast market, a huge investment is required at both locations. At Geneva, the facilities throughout the entire plant would either have to be replaced or upgraded at a cost estimated to be about $1 billion.

At Pittsburg, California, a new cold-reduction mill, with modern pickling and annealing equipment, would cost in the neighborhood of $400 million. Contemplating this outlay and the desire to remain in the western market, United States Steel, in 1986, entered into a joint venture with Pohang Iron and Steel Co., Ltd., of South Korea. The venture involves rebuilding the Pittsburg finishing facilities and bringing in 1 million tons of hot-rolled coils from the new plant that Pohang has just constructed at Gwangyang, South Korea. The Gwangyang plant has the most modern hot-strip mill in the world and produces the highest quality hot-rolled coils. These will be converted on a six-high cold-reduction mill which is under construction at Pittsburg, California. The new facility in California will be owned on a 50-50 basis by United States Steel and Pohang. Using the highest grade of hot-rolled coil and the most modern cold-reduction equipment, the plant will produce steel products that can meet any specifications. Further, the total investment of United States Steel is some $200 million, as opposed to $1.4 billion if it had decided to modernize the Geneva and Pittsburg, California, plants on its own.

Regrettably, jobs have been lost as a result of the closure of the Geneva site. However, it is reasonably certain that, within a short time, the plant would have been abandoned, since its replacement would have been too expensive; thus, the jobs would have been lost anyway. Further, the jobs at Pittsburg, California, are guaranteed by the new joint venture.

The new facilities will be finished in 1989, at which time 1 million tons of hot-rolled coils will be brought in from South Korea. Until that time, the existing facilities at Pittsburgh, California, must be supplied with coils from sources other than the closed Geneva site. The source will be mainly the Fairless works and insures that the plant will be producing extra tonnage of hot-rolled coils at least through the fourth quarter of 1989.

In another area, Fairfield, Alabama, an investment of $750 million was made in a modern, world-class seamless-pipe mill, along with a continuous caster to supply rounds. This is the most modern and, in many respects, the only modern seamless mill in the United States and places United States Steel at the forefront of technology in the production of oil-country tubular goods.

At the termination of the six-month strike on February 1, 1987, an announcement was made that United States Steel, in addition to idling indefinitely the Geneva and Texas works, would install a continuous slab caser at Fairfield, Alabama, and upgrade the hot-strip mill there. Another continuous caster for slabs will be installed at the Edgar Thomson works, while the Irvin works (which is supplied by Edgar Thomson) will have its hot-strip mill upgraded. This assures the production of steel in the Pittsburgh area indefinitely into the future.

Summing up, United States Steel, a wholly owned subsidiary of USX, now has five integrated plants operating, having complately abandoned five other integrated plants and reduced a sixth to an electric-furnace operation. The number of integrated plants will be further reduced if the Fairless works hot end is closed down. The integrated plants of United States Steel will then consist of Gary, Indiana; Lorain, Ohio; Fairfield, Alabama; and the Pittsburgh, Pennsylvania, works.

Weirton

Weirton Steel Corporation was formed in 1984, when National Steel spun off the plant located at Weirton, West Virginia, which had been National Steel's integrated plant for more than half a century. The company decided to spin off Weirton in 1982, and the transaction was completed in January 1984.

Weirton was organized as an employee stock ownership plan (ESOP) and has since been quite successful due to a number of factors. First, the plant was kept in operation until its transfer to the new owners. Second, it had reasonably good production facilities. Third, the plant produced a quality tinplate that is highly acceptable to the trade. Fourth, the employees took a 32 percent cut in wages and froze the wage rate for six years.

The arrangement for the transfer of Weirton from National to the new organization was based on two 10 percent notes, one for $47.2 million and the other for $72 million, the first due in 1993 and the second in 1998. Further, the facilities were sold on the basis of 22 cents on the dollar. Therefore, the depreciation on that basis was less by far than it would have been if the facilities were purchased for 100 cents on the dollar.

Weirton has four blast furnaces but no operating coke ovens, so it must purchase its total coke requirement. The steel is produced in basic-oxygen converters and poured into one of the earliest wide-slab casters installed in the United States. Its finishing facilities range from a 60-year-old hot-strip mill, which has been rejuvenated, to what is presently the most modern cold-reduction mill in the United States. There are definite plans to replace the caster with a new one that will have the capacity to cast all of Weirton's raw steel. The current one can only handle about 40 percent of steel production. This will insure considerable saving. Work will also be done on the hot-strip mill.

In 1987, Weirton received a grant of $52 million from the U.S. Department of Energy to build the facilities necessary for the KR process for making iron. This is a new technological development which bypasses the blast furnace. The award was won in competition with the State of Minnesota to build a unit that upon completion will be capable of producing 350,000 net tons of iron per year. The total cost, including auxiliaries, is estimated at about $123 million. The engineering will require approximately one year from June 1987, when the final contract with the Department of Energy was signed. When the engineering is completed, the actual construction will require a year and a half; thus, the facility will start up in late 1989 or early 1990. Weirton has invited a number of steel companies to participate financially in the project.

Fortunately for Weirton, there is a KR plant under construction in South Africa, which is scheduled to begin operation in October 1987. Observation of this facility will be helpful to Weirton in the construction of its plant.

Weirton will be responsible for raising in one way or another approximately $68 to $70 million of the cost. Part of this will be a credit for work done by Weirton Steel employees.

Weirton's principal product is tinplate, which constitutes over 50 percent of its output. The company has been profitable in three years since its inception, and it appears that this integrated plant will continue to function into the future.

Wheeling-Pittsburgh

The steel operations of Wheeling-Pittsburgh were carried out in two integrated plants, one at Monessen, Pennsylvania, and the other with facilities located at Steubenville, Ohio, and Wheeling, West Virginia.

In 1982 and 1983, the company undertook a major capital investment program which involved the installation of two continuous-casting units, one at Steubenville for slabs and the other at Monessen for blooms to feed the newly constructed rail and structural mill. Both casters went into operation in 1983, as did the new rail mill built with a government loan guarentee. The investment involved in the the new facilities was over $500 million. With the depressed condition in the steel industry, Wheeling-Pittsburgh found itself unable to handle the financial charges; consequently, it filed for chapter 11 in April 1985.

Subsequent to the filing, a strike took place, when the company proposed a wage settlement under chapter 11 of $15 an hour. The strike lasted for 92 days, after which time, a settlement was made reducing employment costs to $18 an hour. In reorganizing the company, the new management saw fit at first to close the iron-and steelmaking facilities at Monessen and feed the rail mill with blooms that had been stocked before the strike and then with purchased blooms. The facilities at Steubenville and Wheeling were kept in operation. As a consequence, the list of integrated steel plants was reduced by one, namely, Monessen. Subsequently, Wheeling-Pittsburgh agreed to allow the government to repossess the rail mill, so that it is completely out of the rail business. This willingness was due to a number of factors, including a drastic drop in rail demand. Thus, the plant at Monessen was completely closed. A number of bidders were interested in purchasing the rail mill; as yet, no disposition has been made of the property. If it is purchased by Bethlehem Steel, blooms will be provided from its Steelton, Pennsylvania, works. If another bidder succeeds in purchasing the plant, it will be necessary to bring blooms in or install an electric furnace that can produce steel to be cast at the plant and then rolled into rails.

The plant at Stuebenville, which is dedicated to the production of light, flat-rolled products, continues in operation and, because of the chapter 11 situation, operates efficiently since some cost elements have been eliminated.

Wisconsin

Wisconsin Steel Corporation, originally a wholly owned subsidiary of International Harvester, was sold in 1977 to Envirodyne Industries. Its principal customer was the International Harvester Corporation which, in 1978 and 1979, purchased a large tonnage, amounting to about 25 percent of the company's shipments. The contract between International Harvester and Environdyne called for the purchase by International Harvester of a substantial amount of steel over a number of years. Unfortunately, in 1979, International Harvester sustained a strike of considerable duration and, therefore, did not purchase the amount of steel that was contemplated when the company was sold.

This had a severe effect on the financial condition of Wisconsin Steel, which was at the time in the midst of a rehabilitation program involving expenditures of some $45 million on a blast furnace, as well as considerable sums for other facilities. The company had a $95 million loan guarantee from the U.S. Department of Commerce, of which it had taken $55 million. The banks moved in, seized the inventory, and forced the company to close its doors in April 1980. Thus, three blast furnaces, a battery of coke ovens, a basic-oxygen steel shop, and several bar mills were eliminated from the steel industry's capacity in the United States. Raw-steel potential at Wisconsin was approximately 1 million tons annually.

The closure was particularly unfortunate since the rehabilitation of the blast furnace would have resulted in one of the most efficient units in the United States. Not only the furnace, but also the stoves and auxiliary system were being completely renovated. It would have been finished within a month after the closure at a cost of approximately $2 to $3 million more. Another unfortunate factor was that Wisconsin Steel had built up a very good reputation with its customers. In fact, before the final closing, a survey was made of its 20 largest customers, and 19 of them agreed to do business with Wisconsin if its operations were resumed.

Conclusion

From the foregoing discussion, it is evident that the integrated mills in the country have declined dramatically from 1975 to 1987. Table 2-1 indicates the decrease in the number of integrated mills from 1975 to 1987.

Table 2–1
Integrated Plants in Operation, 1975, with Number Taken Out, 1975–87

	Number in Operation	*Number Subsequently Taken Out*
Alan Wood (closed)	1	1
Armco	3	1
Bethlehem	5	2
CF&I	1	1
Crucible	1	1
Cyclops	1	1
Ford (now Rouge Steel)	1	0
Inland	1	1
Interlake (now Acme Steel)	1	2
J&L[a]	3	1
Kaiser (closed)	1	0
Lone Star	1	1
McLouth	1	1
National	3	0
Republic[a]	6	2
Sharon	1	0
United States Steel	10	5
Wisconsin Steel (closed)	1	1
Wheeling-Pittsburgh	2	1
Youngstown Sheet & Tube[a]	3	2

[a]J&L, Republic, and Youngstown Sheet & Tube merged into LTV Steel.

In 1975, there were 47 mills; by 1985, the number had been reduced to 23. There are indications that there will be further reductions in the number and, in some instances, in the size of integrated steel plants.

The overcapacity that still exists in the U.S. steel industry despite the reduction to 112 million tons, will require the closure of some basic facilities. This will work a hardship on those employed, but nevertheless, if the present capacity (which includes not only the integrated plants, but 50 minimill plants) remains at current levels, the downward bias it exerts on the price structure will make it difficult for the steel companies to be profitable.

Despite the fact that there will be fewer integrated mills, those remaining will produce more than half of the raw steel made in the United States to the end of the century. The steel industry in the United States will be required to produce some 80 to 90 million tons of raw steel on an annual basis for the next 10 or more years. This can only be accomplished if integrated mills represent a significant segment of the industry. It would be impractical to think of a large number of small electric-furnace plants producing this much raw steel. Further, there would not be enough scrap available. Therefore, integrated mills operating on basic raw materials, capable of producing several million tons each, are necessary if the steel requirements of the U.S. economy are to be met.

The integrated plants that remain after the reduction in capacity will be the most modern, efficient units that the industry has to offer. They will also produce products that are extremely difficult although not impossible, for the small electric-furnace operations to make.

As to the future, there will be no fully integrated mill built in the United States for the next 10 to 15 years. None have been built since Bethlehem's Burns Harbor plant was completed in 1970.

Notes

1. Armco, Inc., *Annual Report*, 1983, p. 5.
2. Kaiser Steel Corporation, *Annual Report*, 1975, p. 11.
3. Kaiser Steel Corporation, *Annual Report*, 1980, p. 2.

3
Minimill Companies

There have been small nonintegrated steel mills functioning in the United States for many years, going back before the turn of the century. For the most part, they consisted of a cold-metal, open-hearth shop; a breakdown mill to reduce the ingots to manageable size; and a bar mill which rolled a variety of bar products, such as concrete reinforcing bar, smooth bar, and small structural members. In the post–World War II period, these plants began to shift to electric furnaces which proved to be more efficient than the open hearths. A number of new small plants were added in the early to mid-1960s, when the name *minimill* was applied.

During the 1960s and 1970s, this section of the steel industry grew rapidly, as mills sprang up in various locations throughout the United States. Currently, depending on the definition, there are 50 minimills throughout the country, operated by 31 companies. The extent of the growth is evident from the fact that in the early 1960s, there were fewer than 15 mills that would meet the definition of a minimill (less than 300,000 tons of steel-melting capacity, electric furnaces, a breakdown mill or a continuous caster, and a bar mill).

Recently, there has been a trend toward consolidation as a number of companies have acquired more than one mill. In the early 1960s, there were few instances of this. As of 1987, seven companies operate more than one plant. These include Florida Steel, Nucor, North Star Steel, Birmingham Steel, Structural Metals, Newport Steel, and Atlantic Steel. Table 3–1 lists minimills by number of furnaces, annual capacity, and location by company.

In terms of production, although more testing is required than previously, specifications for typical minimill products (such as

Table 3–1
Minimill Segment of the U.S. Steel Industry

Company	*Plant Location(s)*	*Number of Furnaces*	*Annual Raw-Steel Capacity (net tons)*
Atlantic Steel Co.	Atlanta, Ga.	2	450,000
	Cartersville, Ga.	1	300,000
Auburn Steel Co.	Auburn, N.Y.	1	400,000
Bayou Steel Corp.	LaPlace, La.	2	650,000
Birmingham Steel Corp.	Birmingham, Ala.	1	200,000
	Chesapeake, Va.	2	125,000
	Emeryville, Calif.	1	160,000
	Jackson, Miss.	1	210,000
	Kankakee, Ill.	2	165,000
	Seattle, Wash.	2	220,000
Border Steel Mills, Inc.	El Paso, Tex.	2	220,000
Calumet Steel Co.	Chicago Heights, Ill.	2	150,000
Cascade Steel Rolling Mills, Inc.	McMinnville, Ore.	2	400,000
Chaparral Steel Co.	Midlothian, Tex.	2	1,500,000
Charter Manufacturing Co., Inc.	Chicago, Ill.	1	130,000
Commercial Metals Co.[a]	Birmingham, Ala.	1	250,000
	Seguin, Tex.	2	400,000
Florida Steel Corp.	Charlotte, N.C.	2	280,000
	Jackson, Tenn.	1	400,000
	Jacksonville, Fla.	1	400,000
	Knoxville, Tenn.	2	225,000
	Tampa, Fla.	2	280,000
Georgetown Steel Corp.	Georgetown, S.C.	2	700,000
Hawaiian West. Steel, Ltd.	Ewa, Hawaii	1	60,000
Hurricane Industries, Inc.[b]	Sealy, Tex.	1	140,000
Keystone Consolidated Industries	Peoria, Ill.	2	700,000
Marion Steel Co.	Marion, Ohio	2	300,000
Milton Manufacturing Co.	Milton, Pa.	3	150,000
New Jersey Steel Corp.	Sayreville, N.J.	2	480,000
Newport Steel Corp.	Ashland, Ky.	2	300,000
	Newport, Ky.	3	550,000
North Star Steel Co.	Beaumont, Tex.	2	800,000
	Monroe, Mich.	1	500,000
	St. Paul, Minn.	2	420,000
	Wilton, Iowa	1	275,000
	Youngstown, Ohio	2	360,000
Nucor Corp.	Darlington, S.C.	5	540,000
	Jewett, Tex.	5	550,000
	Norfolk, Nebr.	5	560,000
	Plymouth, Utah	2	425,000
Owens Electric Steel Co.	Columbia, S.C.	3	100,000
Raritan River Steel Co.	Perth Amboy, N.J.	1	750,000
Razorback Steel Corp.	Newport, Ark.	2	220,000

Table 3-1 continued

Roanoke Electric Steel Corp.	Roanoke, Va.	3	500,000
Seattle Steel, Inc.	Seattle, Wash.	2	500,000
Sheffield Steel Corp.	Sand Springs, Okla.	2	550,000
Steel of West Virginia, Inc.	Huntington, W.Va.	2	300,000
TAMCO	Etiwanda, Calif.	1	300,000
Tennessee Forging Steel Corp.[c]	Harriman, Tenn.	3	200,000
Thomas Steel Corp.	Lemont, Ill.	3	270,000
Total U.S. capacity			19,015,000

Source: Direct survey of minimill companies as of mid-1987.

[a]Company's plants operate as SMI Steel Inc. and Structural Metals, Inc., respectively.

[b]Company's plant is presently shut down with reopeining scheduled for 1988.

[c]Purchased in 1986, the presently idle plant is scheduled to reopen in mid- to late 1987.

rebars, angles, channels, narrow flats, and smooth bars) have not changed to any noticeable extent during the past two decades. Consequently, it is possible for a minimill to make these products and still meet specifications with a facility that is 15 to 20 years old. There is, however, an incentive for the company to improve its facilities because of the need to remain competitive with other minimills in costs and quality.

The mills that have been installed since 1975—and there are a number of them—have reasonably competitive equipment and need very few major improvements. Mills installed before 1975 have, in many instances, been updated in order to compete, not so much with the integrated mills, but among themselves.

Although minimill product lines have been limited, in recent years, there has been a movement by a few minimills to produce some products that were formerly considered the sole province of the integrated mill. These include sheets and seamless pipe. A review of the minimills indicates the changes that have taken place, as well as the interest of some in moving into other products. The basis for this move is the overcapacity to produce the traditional minimill products. Consequently, if the company wishes to grow, it must move into other product lines or acquire more plants.

The principal activity in this regard is the action taken by Chaparral to produce medium structural sections and the proposal by Nucor to build plants that will produce flat-rolled products and medium to large structurals. Further, North Star is constructing a seamless-pipe mill at its Youngstown plant. There has been speculation that more minimills will attempt to produce these products. In order to determine the validity of this speculation and to assess the general health of the minimills, interviews were conducted with 29 minimill companies representing 48 plants.

The interviews indicate that some minimills have been quite successful and profitable during the period of depression for the integrated steel companies. However, a significant number have failed during the past two decades and have either been sold or closed; in fact, several have been sold more than once.

Those that have been sold and are currently operating under new management include:

North Star Steel in Minneapolis, sold to Cargill, but maintains its original name.

North Star Steel of Texas, formerly Georgetown Texas, a part of the Korf organization.

Razorback Steel, in Arkansas, originally a division of Tennessee Forging and subsequently a division of Birmingham Bolt, before it became Razorback.

Knoxville Iron, sold to Azcon Corportion, then to Blue Tee, and, in turn, to Florida Steel.

New Jersey Steel, built by the Italians and sold to VonRoll, a Swiss company.

Cascade Steel, sold in 1972 to the Klinger interests and subsequently, in 1984, to the Schnitzers, large scrap dealers.

California Steel in Chicago, sold to a group and is now Charter Steel.

Connors Steel in Birmingham, Alabama, acquired by H. K. Porter in the late 1940s and, in 1985, sold to Structural Metals.

Steel of West Virginia, once a part of H. K. Porter's organization, subsequently sold to a group in West Virginia.

Marion Steel in Ohio, originally Pollock Steel, acquired by Armco and then sold to the group which is now Marion.

Sheffield Steel in Sand Springs, Oklahoma, formerly a part of Armco.

Kentucky Electric, acquired by Republic Corporation of Los Angeles, and subsequently sold to Newport Steel.

Hunt Steel plant in Youngstown, Ohio, filed for bankruptcy and was bought by North Star Steel.

Birmingham Steel, recently organized, purchased six plants in its three-year history. These are:

1. Birmingham Steel, which was part of Birmingham Bolt. Prior to that it was owned by Ceco and before that was known as Southern Electric.
2. Jones & McKnight at Kankakee, Illinois, formerly part of Birmingham Bolt and prior to that known as Jones & McKnight Steel.
3. Jackson Steel in Mississippi, acquired in 1985.
4. Intercoastal Steel in Virginia, purchased in 1986.
5. Northwest Steel Rolling Mills in Seattle, purchased in 1987.
6. Judson Steel in California, purchased in 1987.

Seattle Steel in Seattle, purchased from Bethlehem Steel.

Bayou Steel near New Orleans, purchased by RSR, a Texas company, from the original owner, Voest-Alpine of Austria.

Plants that have been closed include:

Soule Steel in southern California

Marathon Steel in Arizona

Roblin Steel in Buffalo, New York

Youngstown Steel in Youngstown, Ohio

Witteman Steel in southern California

Interviews with the companies reveal considerable diversity in their current status, present problems, and future plans. A discussion of each company follows. These companies with multiple plants are considered first as a group, after which the single-plant companies are presented.

Florida Steel

Florida Steel Corporation, founded in 1948, operates five mini steel plants with an aggregate capacity of 1.6 million to 1.7 million tons. The plants are located at Tampa and Jacksonville, Florida; Charlotte, North Carolina; and Jackson, Tennessee. Its most recent unit, an acquisition, is at Knoxville, Tennessee. A sixth plant, located at Indiantown, Florida, has been shut down for some time and will probably not be reactivated. In addition to the steelmaking facilities, Florida has 10 plants that fabricate concrete reinforcing bars.

The original plant at Tampa went into operation in 1957, at which time it had a modest capacity of some 51,000 tons. Since then, it has been expanded to 280,000 tons.

The plant at Charlotte, North Carolina, was built in the late 1960s, while both the Jacksonville, Florida, and the Jackson, Tennessee, plants were later additions. The former was put in operation in 1976. The latter, with a 400,000-ton capacity, went into operation in 1981. This modern, greenfield-site addition to the company's facilities has one 120-ton electric furnace, as well as continuous casting and rolling facilities. The plant was constructed at a cost of approximately $60 million. Future plans for this relatively new plant call for upgrading the equipment whenever possible or necessary. In keeping with this concept, the plant will have a new cooling bed in the very near future at a cost of $5 million.

Improvements to the other facilities include a virtual replacement of the facilities at the North Carolina plant, where a new mill and continuous caster have been installed. This enabled the plant to increase its capability by some 76,000 tons. An addition to the rolling mill at Jacksonville enables that plant to produce rods as well as rebar. Future plans for the Tampa plant call for a complete replacement of all of the facilities, to be accomplished gradually through the installation of a new electric furnace and a continuous caster, facilities which will supply billets for the existing rolling mill until a new rolling mill is built. It is estimated that the cost of the new plant will be approximately $60 million.

The plant at Jacksonville has an excellent productivity record. The man hours per ton are as low as 1.1, when production is centered around the heavier bar sections. When no. 3 rebar is rolled, since it is a lighter section, the man hours per ton increase to 1.6 to 1.7.

In respect to employment conditions and costs, with the exception of the North Carolina plant (where the United Steelworkers of America have some activity) the other plants are nonunion. Total employment cost is about $15 an hour. However, an incentive system provides additional income over the base wage.

Imports have been a problem with regard to rebar, whose tonnage has been notably increased. In 1986, 78,482 tons of rebar entered the United States through the port of Miami. Florida Steel has a policy of meeting the prices of imported steel on a selective basis. It meets the price for a particular size rebar in a specific location. For example, if no. 5 rebar comes in in substantial tonnages at Fort Lauderdale, Florida Steel will immediately meet the price of no. 5 rebar at that location. However, this price reduction would not apply to other regions that are far removed. In respect to imports, the company has spearheaded the drive by the Steel Bar Mills Association against imports that were either dumped or subsidized. A suit was filed against Peru and a great deal of discussion was held with the U.S. Trade Representative in respect to other countries. This resulted in a decided curtailment of rebar imports from Taiwan, Brazil, Trinidad, and the Dominican Republic.

For many years, Florida Steel has obtained its supply of scrap to feed the furnaces at its plant locations through the David Joseph Company, an arrangement that relieves Florida of the necessity of dealing with a variety of scrap merchants. According to both David Joseph and Florida, the arrangement has been eminently satisfactory.

Florida Steel, for the most part, makes typical minimill products with about 50 percent of the output in rebar and the remainder spread among channels, angles, narrow flats, smooth bar, small structural sections, and rods. In response to a query as to whether the company would consider entry into the flat-rolled product business, the answer was negative.

Unlike many of the minimills and integrated companies, Florida has been profitable, except for 1982 and 1983. In 1984, 1985, and 1986, profits have ranged from $6.6 million in 1984 to $15.4 million in 1986.[1]

Nucor

Nucor built its first plant at Darlington, South Carolina, in 1969 with a 120,000-ton capacity. Subsequently, it constructed plants in Nebraska in 1974, in Texas in 1975, and in Utah in 1981. The annual capacity of the company in 1974, when only the Darlington and Nebraska mills were operating, stood at 400,000 tons. The addition of the mill in Jewett, Texas, added 200,000 tons to the company's capacity. During the latter part of the 1970s, the mills were expanded so that by 1981, when the Utah plant was added, total capacity was well over 2 million tons.

Currently, Nucor's production consists of standard minimill products with some special-quality bars. The company also operates a Vulcraft division which is the nation's largest producer of steel joist girders. This division has a total of six plants and is a substantial consumer of Nucor's products. Another division operates plants in Nebraska, South Carolina, and Utah for the production of cold-finished steel bars. There are, in addition, bolt-making facilities in Indiana with a capacity to produce 40,000 tons annually. Thus, Nucor is much more than a mini steel mill. Its steel production was almost 2 million tons in 1986.

The company is of interest not only for its past and present performance, which has been outstanding in terms of profitability, but also for its future plans. In terms of past financial performance, the company has been profitable every year during the 1980s. While others have registered large losses, Nucor had significant profits, ranging from $22.2 million in 1982 to $58.5 in 1985.[2]

As far as the future is concerned, Nucor has made an announcement that it will build two plants, one to produce flat-rolled products and a second to produce medium and large structural sections. The flat-rolled products mill will be located in Indiana so that it can take advantage of the superior scrap available in that area as compared with that available in the southeast. This is needed to produce steel sheets of deep-drawing quality, since it will contain a minimum of contaminants. However, there still remains the question as to whether a 100 percent scrap charge will produce quality sheets for a number of applications.

The new mill is an innovation insofar as it will cast thin slabs from electric-furnace steel. The slabs will be no more than 2 inches thick and once cast will pass through a holding furnace to keep the temperature before being reduced to sheets in a 4-high, 4-stand hot-strip mill, 54 inches wide. Since the slabs are thin, a roughing stand is not needed, and they can proceed directly to the finishing stands. Production will be 800,000 tons. Some 400,000 tons of these hot-rolled bands will be cold-reduced on a reversing mill. Nucor plans to use a significant tonnage in its own operations, probably as much as 300,000 tons of cold rolled. Hot rolled will be sold. The entire project is estimated to cost some $225 million.

In addition to the flat-rolled mill, Nucor has entered into a joint venture with Yamato Steel of Japan to construct a mill on the Mississippi River in Arkansas, which will produce some 650,000 tons of structural shapes. Of this tonnage, 550,000 will be wide-flange beams, ranging from 10 inches to 24 inches in width. The other 100,000 tons will be tieplates for railroad rails. The total cost of this mill is estimated at $190 million, of which some $40 million will be debt. The partnership gives Nucor 51 percent and Yamato 49 percent. Nucor has also announced its intention to build a steel-fabricating plant. With the completion of these facilities, Nucor will be in a position to offer a number of products to the construction industry, making it a significant force in this area.

Nucor's mills are nonunion. The base pay is relatively low at $8.50 an hour. However, there are incentives which amount to at least an additional $8.50. With fringe benefits, the total employment cost is about $22.00 an hour. In lieu of a pension plan, the company puts 10 percent of pretax profits into a fund which the employees share, and at retirement, they can take a lump sum or annual amounts. Some have as much as $100,000 in this fund. During most of its existence, Nucor has not laid off employees, although at times the work week has been reduced to four days.

The two new plants that will soon be under construction by no means fit the description of minimills in terms of either product or tonnage, although they are electric-furnace operations, as contrasted with integrated blast-furnace and basic-oxygen steelmaking operations. The flat-rolled products plant will be watched carefully by steel industry people all over the world. The plant constitutes an innovation, and, if successful, could draw one or two other small, nonintegrated companies into the production of flat-rolled products. However, the investment is large and this will deter most of the small nonintegrated plants.

In terms of market, Nucor's South Carolina mill covers the Southeast, the Texas mill covers Texas and Louisiana as well as Oklahoma, the Utah plant reaches as far as California (its major market), and the Nebraska mill reaches as far as Denver.

North Star

North Star Steel Corporation, a subsidiary of Cargill, the agricultural giant, entered the steel business in 1974. The first plant acquired was the North Star Steel Company of Minneapolis, which was owned in part by Co-Steel Company of Canada with Cargill holding a significant interest. Cargill bought out the other partners in 1974 and became the sole owner of a plant built in 1965 with a capacity of 350,000 tons.

Since its entrance into the steel industry, Cargill, through North Star, has acquired five additional plants with a total steelmaking capacity well in excess of 2 million tons.

In 1977, North Star acquired a plant in Wilton, Iowa, which a group of local citizens had under construction. After the acquisition, North Star finished the plant with its capacity of 250,000 tons.

In 1980, the company constructed a plant at Monroe, Michigan, with a capacity of 400,000 tons. It was intended to be a special-quality–bar producer. However, because of some problems, it did not develop as expected. Changes made recently, including improvements to the bar mill, will allow the plant to produce 90 percent of its output in special-quality bars as opposed to 40 percent.

In 1983, Georgetown Steel of Texas was acquired. It is one of the most modern rod mills in operation today. North Star paid $1.00 for the plant and assumed the company's debts, which were settled with the banks for about 50 cents on the dollar. Capacity at the Texas plant, known as North Star Texas, is rated at 700,000 tons. However, the two 120-ton furnaces, if pushed, could produce significantly more.

In 1985, the company continued its acquisition policy by purchasing the bankrupt plant of Hunt Steel in Youngstown, Ohio, for $22 million. This facility consisted of two electric furnaces (bought second-hand in England), a new continuous caster, and a seamless-pipe mill which was not functional. The plant's capacity is listed at 300,000 tons.

In 1986, Ohio River Steel at Carson City, Kentucky (another bankrupt company) was acquired and its name changed to North Star Kentucky. This plant, which consists of a structural mill and no steelmaking facilities, must be supplied with purchased semifinished steel in the form of blooms and billets. However, by 1988, adjustments made to the continuous caster at Youngstown will allow it to produce blooms for the Kentucky plant, while some 150,000 tons of billets will be provided by the St. Paul mill. Perhaps North Star's most significant undertaking is the construction of a seamless-pipe mill at the Youngstown plant at a cost of $90 to $100 million.

At present, the seamless-pipe market is badly depressed. Thus, the question arises as to the wisdom of installing a new seamless-pipe mill in the midst of a depression in oil-country tubular goods. As of 1987, the number of rigs drilling for oil in the United States is less than 800, as compared with 5,430 in the boom year of 1981. The demand for oil-country tubular goods is dependent on the number of rigs operating. Those well-informed about the oil industry estimate that the price of oil will have to rise to $25 a barrel before an appreciable number of rigs (over 2500) will be in operation.

When North Star purchased the Youngstown facility and announced its intention to build a seamless-pipe mill, the LTV Steel Corporation offered its mill located nearby to North Star. The facility had been updated in the early 1980s, with an investment of some $70 million, and LTV considered it a state-of-the-art operation. North Star, however, preferred to proceed with its own installation. As presently planned, this mill will not have the ability to finish oil-country tubular goods, so another facility was required to take the unfinished pipe and perform finishing operations. In spring 1987, North Star acquired an 80 percent interest in Universal Tubular Services, a company in Houston, which will finish the pipe made at Youngstown by threading and heat-treating it. The capacity is about 300,000 tons, enough to take care of the entire production from the Youngstown mill. The pipe mill at Youngstown will be placed in operation in December 1987.

When the mill is in operation, North Star is convinced that its only competition will be the new United States Steel seamless-pipe mill recently installed at Fairfield, Alabama. This plant has the capacity to produce 600,000 tons of oil-country tubular goods and was built along with the continuous caster. The North Star plant at Youngstown is being financed by Cargill, which will take a 75 percent equity interest.

In terms of improvements at its other plants, only relatively minor additions and modifications will be made, so that these mills will be able to produce a certain amount of special-quality bars. Some speed controls, as well as improvements in reheating furnaces, will be required. In general, there is very little need to upgrade the mills themselves, since most of them are relatively new, and the specifications for their products have not been significantly increased.

The company has substantial backing from its parent, Cargill. However, it is not an innovator, but rather content to install equipment and operate in those areas where experience has proven that profits can be made. In line with this philosophy, there is interest in the flat-rolled–products operation underway at Nucor. If this is successful, it is quite possible that North Star will install a light, flat-rolled–products operation. However, this will not take place until the early 1990s.

Several of North Star's plants, including Iowa, Michigan, Texas, and Youngstown, are nonunion, with wages equal to those paid by the major mills. However, there are not the same fringe benefits, so that the total employment cost per hour is considerably less than $20. There is, however, a profit-sharing plan which augments the income of the employees.

Birmingham Steel

Birmingham Steel Corporation is a recent entry into the steel business at the minimill level. It was founded by AEA Investors, Inc., a New York firm, and began operations through the acquisition in August 1984 of two steel plants. These facilities, located in Birmingham, Alabama, and Kankakee, Illinois, were operated by Birmingham Bolt Company. The acquisition represented a $40 million investment.

In 1985, Birmingham Steel, for $40 million, acquired the Mississippi Division of Magna Corporation, the only minimill in Mississippi. This acquisition almost doubled the capacity of Birmingham from 240,000 tons to 470,000 tons.[3] In 1986, Birmingham acquired Intercoastal Steel Corporation of Norfolk, Virginia, for $6.5 million, thus adding a fourth mill to its system. The addition of this company, as well as improvements made at the other plants, particularly at Birmingham and Kankakee, increased the capacity to 700,000 tons in 1987. Early in the same year, Birmingham acquired two other facilities, the Northwest Rolling Mill Company in Seattle and the Judson Steel Company near San Francisco, thus lifting the capacity to approximately 1 million tons.

The two furnaces at the Birmingham plant, when it was acquired, were small, rated at 9 tons and 11 tons per heat. These were replaced by a 50-ton furnace, with a 32.5 KVA transformer, enabling the operators to produce a heat in 75 minutes tap-to-tap time. Mississippi has a 50-ton furnace, Kankakee has two 20-ton furnaces, and Norfolk has two 35-ton furnaces.

The company's output, for the most part, is restricted to concrete reinforcing bars, a large percentage of which are processed into mine-roof bolts. Birmingham is the largest producer of this product in the United States, accounting for approximately 50 percent of the roof bolts and some 70 percent of the roof plates used in the domestic underground coal-mining industry. In addition to the six steel mills, Birmingham has ten plants producing roof bolts and plates. These facilities are located near the major underground coal-mining areas.

In the short space of 2 1/2 years, Birmingham Steel grew from nothing to a 1-million-ton company. In 1986, shipments amounted to 436,000 tons, of which 65 percent were sold to outside consumers, while the remainder was used internally for roof bolt and plate production.

In terms of capital investment, the company expects to spend $25 million on improvements, of which approximately one-half was spent in 1986. The mill at Kankakee was the recipient of a continuous-casting unit which allowed it to resume operations in May 1986 after an 11-month shutdown. It is estimated that the investment there will save approximately $30 a ton. The installation of the 50-ton furnace at the Birmingham location, along with improvements in the cooling bed, is estimated to reduce costs by $15 per ton.[4] Very little was required for the Jackson, Mississippi, mill, since it has relatively modern equipment.

The strategy of Birmingham Steel and its philosophy are stated in the 1986 Annual Report:

> Our strategy is to purchase undervalued steel-producing assets and operate them cost-effectively by implementing capital improvements, applying efficient production techniques, and reducing management and labor costs through innovative programs. . . . As we enter fiscal 1987, the company is ideally postured with sufficient capital resources to acquire additional undervalued steel assets and to take advantage of market share opportunities or favorable trends in steel pricing and demand.[5]

Profits for fiscal 1986, a year in which many of the steel companies incurred substantial losses, amounted to $7.9 million. The company has achieved significant reduction in cost by adopting a policy of long production runs of the same product, thus reducing the need to change rolls—a time-consuming operation. In some instances, the run on a particular product may be seven to eight days. The 1986 annual report gives the average cost of producing steel billets at $161 a ton.[6]

Birmingham Steel is a nonunion operation which provides an attractive incentive wage plan for its employees. Over and above the basic hourly employment cost, the plan consists of contributing $12 per ton for every ton above the established level, which is usually set at about half of the facility's capacity. The sum is divided among the personnel engaged at the particular facility. This plan has enabled the employees to attain a wage level that is virtually equal to that paid by the integrated steel companies organized by the United Steelworkers of America.

In 1986 Birmingham had no intention of expanding its business into the light, flat-rolled–product sector of the steel industry. An offer made by a foreign steel producer to form such a venture was turned down when it was estimated that the cost would be over $200 million. The company has found a niche with rebar and, at present, expects to hold to that product, as well as mine-roof bolts, with some small amounts of other products such as angles and channels. It plans to continue to be "a low-cost supplier of reinforcing bar (rebar) to the construction industry and is the nation's largest manufacturer of steel roof-support systems used in the underground mines.[7]

In June 1987, James A. Todd, Jr., Chief Executive Officer of Birmingham Steel, indicated some interest in light, flat-rolled products, now that the technology was readily available. He stated in his speech given on June 24th in New York City that it might be possible to install a thin-slab caster and a hot-strip mill on the basis of a joint venture. The joint venture was stressed particularly in view of the fact that very few minimill companies could afford to make such an investment alone. However, he states that such a project could not be realized before the early 1990s.

Structural Metals

Located in Seguin, Texas, Structural Metals Inc. (now a subsidiary of Commercial Metals) was founded in 1947 with a rerolling mill and no steelmaking capacity. In the mid-1960s, two small electric furnaces were installed, each with a heat capacity of 25 tons. In 1969, the two 25-ton furnaces were replaced with one 50-ton furnace. In subsequent years, improvements were made to the rolling mill. In 1983, a major step forward was taken with the installation of a 90-ton electric furnace while the 50-ton furnace is kept in standby condition. Total capacity for the plant is in excess of 400,000 tons. Products consist of reinforcing bars, angles, merchant bars, squares, fence posts, plain rounds, and some special sections.

The rolling mill is an in-line operation, having been converted from a partial cross-country mill, giving the plant a much more competitive situation in terms of quality and costs. In addition, an improved cooling bed has been installed.

In 1983, Structural Metals purchased the Connors Steel plant of H.K. Porter in Birmingham, Alabama (which had been shut in mid-1983) for $6 million. The melting shop was put back in operation in March 1984, and the rolling mill started up in April of the same year. A considerable capital investment was required to make the plant competitive. The rolling mills have been improved and modernized with the introduction of new stands. The finishing end has been improved, as has the warehouse section of the plant. Reheating furnaces have been upgraded, and the final step in the modernization process will be the revamping of the melt shop in the not too distant future. This may require new furnaces to replace the two 45-ton electric furnaces which have a capacity of 250,000 tons. The plant is also equipped with two continuous-casting machines.

Neither plant has a union. However, in addition to the base pay, there are incentives which add to the employees' income.

The company has expressed interest in the flat-rolled segment of the steel business. However, it has no current plans to enter.

Atlantic Steel

Atlantic Steel Company has been in existence for over eighty years, having been established in 1901. From its start until the 1950s, the company operated open hearths with an ingot practice feeding a bar mill. In 1951, the first electric furnace was installed at the Atlanta plant, followed in 1955 by a second furnace, and, in the same year, a new combination bar and rod mill. By the 1960s, the open hearths had been abandoned, and the plant operated electric furnaces exclusively.

A transformation took place at the Atlanta mill in 1980, when a new Morgan rod mill and continuous caster were installed. The new mill has a finishing speed of 20,000 feet per minute, although this has not yet been attained. Thirty-five percent of the output of the Atlanta plant is in the form of rods, some of which are used internally to manufacture wire, while others are sold on the open market. The bar mill and the rod mill constitute the finishing facilities of the Atlanta plant, producing a variety of products including rods, special-quality bars, and flat sections up to 10 inches wide.

In 1975, a second plant at Cartersville, Georgia, was put into operation with an electric furnace, a continuous caster, and a bar mill, producing rounds, flats, squares, and rebar. The Cartersville plant has a capacity to produce 300,000 tons of raw steel, while the Atlanta plant can produce 450,000, making a total of 750,000 tons.

In 1979, the company was acquired by Ivaco, Inc., a Canadian steel company. At present, the plant is organized by the United Steelworkers of America. However, employment costs are less than those incurred by the integrated mills. Wages of the employees are supplemented by incentive plans in some areas of the plants.

The company has no interest in producing light, flat-rolled steel products.

Newport Steel

Newport Steel Corporation in Newport, Kentucky, had its beginnings in 1909. For years, it was a subsidiary of Merit-Chapman and Scott Corporation and, in 1956, was sold to Acme Steel, before Acme merged with Interlake Iron to form Interlake Steel Corporation. When acquired by Acme, it operated electric-arc furnaces and open-hearth furnaces, with a total steel capacity of 700,000 tons, as well as facilities to produce sheets and pipe. In 1965, the open hearths were eliminated, and the plant operated three electric furnaces.

In July 1980, at the termination of a contract with the labor union, Interlake asked the union to consider a one-year freeze on wages and benefits, asserting that it could not operate profitably if there were an increase in employment costs. The union voted the proposition down and Interlake closed the plant in July 1980.

In early 1981, the present management purchased the plant and started it up again. Its products had been principally pipe, with a limited quantity of sheets. However, during Interlake's last few years, the sheets had not been profitable but were made to satisfy customer demand. The new management quickly noted that sheets were unprofitable and never made them as a product, since there were some quality problems which would have resulted in significant losses. It now produces welded pipe almost exclusively. The size ranges are from 4 1/2 inches to 13 inches in diameter. The capacity of the plant, with three electric furnaces rated at 100 tons each, is approximately 500,000 tons annually.

Kentucky Electric, now a subsidiary of Newport, was established as a steel company in the mid-1960s by the Mansbach family, which also owned a large scrap business. At first, it operated a small 15-ton furnace. Two more furnaces were added in the 1970s. In the late 1970s, the mill was sold to Republic Corporation of Los Angeles, which had been Republic Motion Picture Company. In 1982, the melting capacity was considerably increased as two 40-ton electrics replaced the three smaller furnaces, and capacity was increased from 140,000 to 240,000 tons. In April 1985, there was a strike at the plant, and Republic announced that, without a settlement to their liking, the plant would be closed. This happened in August 1985.

Newport Steel bought the plant in 1986 and actually began operations in October of that year. Currently, the furnaces are producing about 45 tons per heat, giving the plant a 300,000-ton capacity. The output in terms of products is relatively narrow flat sections for spring steel.

Both the Newport plant and the Kentucky Electric plant are organized by the United Steelworkers of America. However, the plants do not pay the rates of the integrated producers. The income of the employees is supplemented by an incentive plan.

The company has no interest in producing light, flat-rolled products.

Auburn Steel

Auburn Steel, located in Auburn, New York, began operations in 1976 with a 150,000-ton capacity for the production of raw steel. It was unusual insofar as the company was completely owned by the Japanese via a combination of Ataka Trading Company and Kyoei Steel Company.

Within a few years, the size of the furnaces was increased from 15 feet to 16 feet, which allowed the production of 70 tons per heat as opposed to 58 tons. Further, the transfomer size was increased from 27,000 KV to 33,000 KV. This reduced heat time from over 2 hours to 1 hour and 20 minutes. As a consequence, the capacity is now 400,000 tons.

Currently, improvements are being made in the reheat furnace and the rolling mill, where a reversing rougher is being replaced with six stands in-line, thus making a decided improvement in the use of rolling technology.

Output consists of the typical minimill products, such as rebar, rounds, squares, narrow flats, channels, and some angles. Some 10 percent of the production is in special-quality bars. Man-hours are somewhat above 2 hours per ton.

The plant is nonunion with all of the personnel on salary. Earnings are supplemented by a profit-sharing plan which has operated successfully. The ownership of the plant changed in 1984 and now is in the hands of Sumitomo Trading Company and Kyoei Steel, both of Japan.

Conversations with the current management indicate it has no intention of entering the flat-rolled segment of the steel business.

Bayou Steel

Bayou Steel Corporation is located at LaPlace, Louisiana, some 35 miles from New Orleans. The plant was built in 1978 by the engineering division of Voest-Alpine, the Austrian steel complex. In addition to the production of steel, the plant was intended to be a showplace to exhibit the technology developed by Voest-Alpine. The intention was to sell steel mill equipment in the United States; however, the decline in steel activity rendered the plan impractical.

From its inception, Bayou Steel suffered substantial losses. In 1986, the plant was sold to RSR Steel Corporation of Texas for $76 million.

The facilities consist of 2 high-powered, 70-ton electric furnaces, which went into operation in late 1981. They are equipped with water-cooled side panels and roofs as well as oxy-fuel burners and are capable of producing a heat in less than two hours. In addition to the furnaces, there are two continuous-casting machines with four strands each. The rolling mill, which was built by Danieli, consists of 15 in-line stands and is one of the most modern in the minimill sector. It is capable of rolling 450,000 tons of products which include small structural sections and beams up to 4 inches wide. The melting capacity is 650,000 tons. The original intent, which did not materialize, was to sell up to 200,000 tons of billets in the Caribbean area. The new ownership intends to maintain the same proportions, rolling 450,000 tons and having the remaining tons available for sale as billets.

The mill is located on water and, therefore, can take advantage of water transportation to bring in scrap and ship out products as far as Pittsburgh and Chicago, where it has mill depots.

The intention of the management is to continue to produce small and medium structural shapes. There is no interest in the light, flat-rolled segment of the steel industry.

In March 1987, a union election certified the United Steelworkers of America as the bargaining agent. The contract calls for employment costs at $17 per hour, with some minor increases through 1990. There are a minimum of work rules and job classifications that will allow the management flexibility in the use of the work force. The contract is to run for six years, which will give considerable stability to the operation.

Border Steel

Border Steel Mills Inc., located in El Paso, Texas, was founded in 1962 with one 20-ton electric furnace. Since that time, it has increased its capacity and currently operates two 35-ton furnaces with an annual potential output of 220,000 tons. In addition, there is a combination cross-country and straightaway bar mill. The output is concentrated on rebars, but there are also special-quality and forging-quality steel products, as well as grinding balls and grinding rods. Scrap is obtained from 2 wholly owned companies which supply most of the mill's requirements.

The management feels that the company has found a particular niche. As a consequence, it has no interest in entry into the light, flat-rolled segment of the steel industry.

The United Steelworkers have the plant organized, although the employment cost per hour is below that of the integrated mills.

Calumet Steel

Calumet Steel in Chicago Heights, Illinois, was a subsidiary of Borg-Warner, along with Ingersoll Steel Division at Newcastle, Indiana, as well as the Franklin Steel Division in Franklin, Pennsylvania. Calumet and Newcastle had electric furnaces, while Franklin was a rerolling operation. In 1985, both Franklin and Calumet became independent, Newcastle having been closed.

Currently, Calumet operates two 30-ton electric furnaces, a continuous caster, and a bar mill, producing minimill products, such as merchant bars, light shapes, and structurals. Capacity is approximately 150,000 tons.

The plant is organized by the United Steelworkers of America. However, the employment costs are below those of the integrated steel plants.

The company has no interest in entering the light, flat-rolled segment of the steel industry.

Cascade Steel

Cascade Steel Rolling Mills is located in McMinnville, Oregon, near Portland. The company was founded in 1968 and began production the following year, with all of the major production facilities consisting of used equipment. These included two relatively small furnaces, a continuous caster (bought from Wickwire Spencer through Keystone Consolidated Industries of Peoria, Illinois), and a rolling mill purchased from Allegheny-Ludlum. At its inception, capacity was approximately 100,000 tons.

There were production problems at the start, and within two years, the company was in financial trouble. Some Portland investors put approximately $2 million into the operation to rescue it. Unfortunately, the money was invested little by little and did not solve the problem. During this period, the cash position was so bad that each truckload of scrap had to be paid for on arrival.

In 1972, the Klinger interests bought the company and were fortunate insofar as the market turned around in the next year. For two years, they enjoyed an unprecedented boom. The Klinger ownership continued until 1984, when the Schnitzer interests, one of the largest scrap operations in the Northwest, bought the company.

During the Klinger and Schnitzer regimes, much of the equipment was replaced. Steelmaking capacity has been significantly increased and, in 1980, a new continuous caster was installed as well as a new reheat furnace, while the electric furnaces were computerized. In 1985 and 1986, a completely new rolling mill was installed, which among other things allows for a very rapid roll change. This can be accomplished in minutes, whereas formerly it took hours. As a consequence of these improvements, the company now has a capacity to produce 400,000 tons and is competitive in the production of rebars, fence posts, narrow flats, and merchant bars.

The plant has the United Steelworkers of America as a union. However, the base pay is lower than that of the integrated producers. It is supplemented by an incentive plan, so the actual take-home pay is not far less than that of the steelworkers in the major integrated companies.

The company has no interest in producing flat-rolled products.

Chaparral

Chaparral, in Midlothian, Texas, was founded in 1975, with a 400,000-ton capacity. The facilities consisted of one 130-ton electric furnace, a continuous caster, and a 15-stand, in-line merchant bar mill. Its size and product mix for the first 6 years of its existence justified placing it in the minimill category.

In 1981, an expansion program was put into effect, which in terms of size and product mix removed the plant from the minimill category. The program consisted of installing an additional 130-ton furnace, a continuous caster, and a structural mill capable of producing wide-flange beams up to 10 inches wide. Subsequent alterations to the furnaces increased the heat size to 150 tons, and the structural beam mill was altered to make it capable of producing 14-inch–wide flange beams. Capacity was increased to 1.4 million tons. The improvements to the equipment consisted of installing water-cooled panels and roofs on the electric furnaces and providing 85,000 KV transformers. The planned installation of a ladle furnace will increase output and enable the company to ship 1.5 million tons of product in the near future.

At present, the man-hour-per-ton output is 1.6 man hours. The goal to be striven for, which seems to be a possibility, is 1 man hour per ton.

Approximately 35 percent of the plant's output is in wide-flange beams, with additional standard "I" beams. The bar mill has a 600,000-ton capacity and produces 300,000 tons of rebar as well as some narrow flats, channels, and universal plates. Further, there are some 70,000 to 80,000 tons of special-quality bars produced. The continuous caster casts "dog-bone" shapes for the structural beams. Currently, there are 11 strands of continuous-casting capacity, and there will be 2 ladle furnaces. Another ambition of the management is to reduce the cost of wide-flange beams close to the level of rebar costs.

The plant is not unionized. The entire work force is on a salary basis. In addition, there is a profit-sharing plan, so that in a good year the labor force could make $25 per hour.

Future plans, after the 1.5-million-ton shipment rate is reached, could call for a new site. This, as yet, is not definite. The company is watching with interest the light, flat-rolled product development that Nucor has undertaken. However, at present, there is no plan for entering that segment of the steel business.

Charter Electric Melting

In 1978, Charter Manufacturing Company bought a bar mill from Great Lakes Steel, relocated it, and began to roll steel bars from purchased billets. In 1981, the company bought the facilities of the California Steel Company, a minimill located in Chicago. The facilities included a 45-ton electric furnace, a continuous caster, and a bar mill. Charter abandoned the bar mill and installed a single-strand, Morgan No-twist rod mill, and Stelmore controlled cooling in the plant in Milwaukee. Thus, it established two plants. In Chicago, steel was melted and billets cast; in Milwaukee, a thoroughly modern rod mill and wire-drawing facilities produced the finished product.

The output is approximately 120,000 tons per year, none of which include traditional minimill products. The company markets wire rods for cold-heading wire.

Since its establishment, Charter has made a niche for itself in the steel business. The Chicago plant—in addition to its electric furnace and continuous caster (which produces 4-inch-by- 4-inch billets)—has ladle metallurgy and argon stirring. Because of the sophisticated products, it is necessary to use select scrap such as no. 1 bushlings with very low residuals, which is available in the Chicago area. The Chicago plant is organized by the Teamsters, but there is no union in the Milwaukee plant.

The company has no interest in entering the light, flat-rolled products area, since it feels quite satisfied with its current product mix.

Georgetown Steel

Georgetown Steel was founded with German ownership in 1969 as an electric-furnace operation producing wire rods. Located in Georgetown, South Carolina, at the outset it had three 60-ton electric-arc furnaces, two continuous-casting machines, and a combination bar and rod mill. Annual capacity was 350,000 tons. Shortly after its founding, the Korf organization built a facility to produce some 350,000 tons of direct-reduced iron pellets. This involved the Midrex Process which depends on gas. The direct-reduced pellets were to be charged into the electric furnaces along with scrap to produce steel for the rod and bar mill.

In 1974, a new rod mill was installed which increased production and improved quality considerably. At the present time, there are two 70-ton electric furnaces, continuous-casting equipment, and a modern rod mill which is state-of-the-art technology. The furnaces are fed with 80 percent scrap and 20 percent direct-reduced iron which is virtually free of impurities. As a consequence, the impurities in the scrap are diluted, and it is not necessary for Georgetown to reach out great distances to secure acceptable scrap.

Production at the present time is almost exclusively rods with an occasional tonnage of rebar, which most probably would not reach 5,000 tons in a year. Total production of continuously cast steel in 1986 came to 640,000 tons.

In 1984, the situation was further improved by the addition of ladle metallurgy, at a cost of $1 million, which provides purer steel. In the past few years, particularly in 1981 and 1982, considerable improvements were installed, including new reheat furnaces, water-cooled panels on the electric furnaces, and a number of changes in the rod mill. Total investment was approximately $30 million. Georgetown will continue to produce rods and, in the future, it may produce some high-quality bar.

The company is organized by the United Steelworkers of America. However, it pays a lower rate than the integrated mills, although this is supplemented by an incentive system.

In terms of future activity, there is an interest but no definite plans to enter the flat-rolled segment of the steel industry. If Georgetown makes such a decision, its use of direct reduced iron in the electric furnace charge produces a steel that is acceptable for the production of sheets.

Keystone

Keystone Steel and Wire Company is a division of Keystone Consolidated Industries with the main headquarters in Dallas, Texas, and the plant located in Peoria, Illinois. The company is essentially a wire producer, since most of its rods are consumed internally, although 80,000 tons are sold on the open market. The two 170-ton electric furnaces, equipped with water-cooled panels, are capable of producing 700,000 tons of raw steel. Currently, this is cast into 3-inch billets and rolled on a rod mill which will soon be replaced.

The new rod mill will be at the cutting edge of technology. Capable of finishing speeds of 20,000 feet per minute, it will also produce a 3,400-pound coil. The caster will be upgraded to handle a 4 1/2-inch billet; it is expected that the quality will be excellent and the cost low. Production from mill will be in the area of 600,000 tons of rods.

Keystone is organized by an independent union. Wages are supplemented by incentive plans based on productivity and profit sharing. In 1986, each person received $1,000 from the profit-sharing program.

Recently, the company purchased three wire companies, so that more of its production will be consumed internally. Having established a base in the wire business, Keystone has no interest in entering the light, flat-rolled products segment of the industry.

Marion Steel

Marion Steel Company is located in Marion, Ohio. In the course of its history, it has had three owners. It was organized as Pollak Steel Company in 1911, but the plant was not built until 1916. For many years, it operated as a rerolling mill with no steelmaking capacity. In the late 1960s, an electric furnace was installed with an annual capacity of 115,000 tons.

In the early 1970s, the company was acquired by Armco, which operated it until 1981, during which time a second electric furnace was added. The plant was shut down in 1981 and very shortly thereafter was sold to a new group which organized it as Marion Steel and began operations in February 1982. Unfortunately, the steel market was depressed at the time, and in September 1983, the new company was forced into chapter 11. Reorganization and emergence from this situation was rapid. By February 1985, Marion moved out of chapter 11.

Currently, the plant has a capacity of 300,000 tons of raw steel annually which is produced in two 50-ton electric furnaces and poured through a 4-strand caster which was recently rebuilt. The bar mill is a cross-country mill which in June 1987 was improved through a new reheating furnace and a new cooling bed. The output of the plant is the usual minimill product range, consisting of rebar, angles, smooth rounds, squares, and narrow flats. Sign posts are also produced.

The work force is not unionized. However, in addition to base pay, the employees participate in a profit-sharing program which assigns 10 percent of the gross profits for distribution.

Marion Steel will continue to produce its current line of products with no interest in entering the light, flat-rolled segment of the steel industry.

Milton Manufacturing

Milton Manufacturing Company is located in Milton, Pennsylvania, north of Harrisburg. The company has three 22-ton furnaces, with approximately 150,000 tons of capacity. In addition, there is a continuous caster and a bar mill. The rebar products for use in highway bridge construction are coated with epoxy to prevent the moisture that seeps through the concrete from rusting the rebars. In order to coat the product effectively, a smooth surface is necessary, which comes from the continuous caster, as well as the addition of vanadium.

The company, given its capacity and its product mix, has no interest in moving into light, flat-rolled products. It has established a niche for itself.

Although organized by the United Steelworkers of America, it has a lower employment cost than the integrated mills. In addition, incentives make it possible for the employees to increase their income.

New Jersey Steel

New Jersey Steel Corporation is a relative newcomer to the minimill sector of the steel industry. Steel was first melted at the plant in Sayreville, New Jersey, in May 1973, and finished products were produced several weeks later. Originally, the plant consisted of a 52-ton electric furnace, a 5-strand casting machine, and a bar mill; all supplied by Italian firms, since the owner of the plant was Aldo Alliata of Italy. The capacity of the original plant was 150,000 tons, but in 1974, it was decided to increase this to 250,000 tons by replacing the electric furnace with a larger one.

In 1977, the ownership of the mill changed hands, as New Jersey Steel was acquired by VonRoll of Switzerland. The plant subsequently beset by labor problems was shut down and not restarted until the end of 1980. Only the rolling mill was in operation at that time, since the melting and casting functions were not resumed until late in 1981. Subsequent to that, a number of improvements were made to the physical assets that put the plant in condition to produce steel competitively. A new management team was brought in at the end of 1982. Since that time, production has increased, reaching a high of 335,000 tons in 1986.

The company produces the standard minimill products, with an emphasis on concrete reinforcing bar. Its sales of this product were given a significant lift when Bethlehem Steel, at its Steelton plant near Harrisburg, Pennsylvania, dropped out of the rebar business. This left a considerable void, since Bethlehem had the capacity at Steelton to produce 600,000 tons of rebar.

At present, New Jersey Steel can roll more than it can melt. Thus, it purchases several thousand tons of billets per month.

The intention of the company is to spend $8 million to increase its capacity from 360,000 tons to 480,000 tons through the installation of a new furnace and a powerful transformer. This will assure that the company will maintain its niche in the rebar business and with other typical minimill products that it will continue to produce.

There is no interest in entering the light, flat-rolled segment of the steel business.

In terms of labor, the shop has been nonunion since 1981. It has a profit-sharing plan in addition to its hourly wage cost.

Owens Electric Steel

Owens Electric Steel Company, founded in 1961, is a part of a larger steel operation which includes fabricating plants and the production of joists. The company operates three furnaces of 18-, 21-, and 25-ton heat sizes, as well as a 3-strand continuous caster and a bar mill. Its capacity has been stated at 100,000 tons. Its products are rebars, angles, and smooth bars.

The company is not unionized and has no practical interest in entering the light, flat-rolled segment of the steel business.

Raritan River Steel

Construction was started on the plant at Raritan River in New Jersey in 1978 and was completed by mid-1980. The first full year of production was in 1981. The mill consists of 1 electric furnace with 150-ton capacity, equipped with water-cooled side panels and a water-cooled roof. Tap-to-tap time on the furnace is 90 minutes. In addition, there is a continuous caster and a 2-strand Morgan rod mill. This mill is the equivalent of any in the world. In fact, there are only a few others that compare with it. It has an exit speed of 20,000 feet per minute, and the weight of the bundle is 4,400 pounds, or 2 metric tons. It is faster than most rod mills in the United States and produces a heavier bundle. The yield of rods from billet is better than 96 percent.

Since Raritan River produces rods exclusively, it is not strictly speaking in the minimill category, for it does not produce any of the typical minimill products. However, many consider it a minimill; thus, it is included here. Raritan is the leading producer of rods in the United States, with some 720,000 tons in 1985.

In addition to the billets produced by the continuous caster, there are 150,000 tons purchased, almost exclusively from Europe. These come into the United States at the port of Newark and are brought to the plant by truck. A large quantity of the rods produced at Raritan River are shipped by truck to Philadelphia, where they are placed on barges and distributed throughout the eastern United States.

The company's market, particularly in highly specialized rods, extends to thirty-six states. Because of the quality demanded in the rods, a great deal of attention must be paid to the scrap that is used. It has to be low in residuals to produce quality rods. Consequently, the selection must be carefully made.

Raritan River is owned by CoSteel of Canada and is nonunion. However, the wage scale is augmented by a profit-sharing plan to the extent of 8 1/2 percent of pretax profits.

The company has no interest in entering the flat-rolled segment of the steel industry.

Razorback Steel

Razorback Steel Corporation came into existence as a division of Tennessee Forging in the late 1960s. Subsequently, it was acquired by Birmingham Bolt, when its production was principally in bars, as well as concrete reinforcing bar. In the late 1970s, the name was changed to Newport Steel, with some accompanying management changes. In 1981, ownership changed again and the firm was named Razorback Steel Corporation. For the better part of two years, it operated its 220,000-ton capacity to produce standard minimill products. This proved unprofitable and, in late 1982, the minimill product approach was abandoned, when it was decided to devote the entire output to railroad tieplates.

In 1983, the mill went into production and was reasonably successful, since there were only two other locations where tieplates were manufactured in the United States. Both of these were plants of Bethlehem Steel, one at Steelton, Pennsylvania, the other in Seattle, Washington. In 1984 and 1985, foreign competition appeared with a much lower price and made it more difficult for Razorback to be profitable. At the present time, with the railroad market considerably depressed, the company is forced to sell at a low price, but it is still profitable.

In addition to its two 30-ton electric furnaces, there is a continuous caster as well as a merchant mill to produce tieplates. The management has no interest in entering the flat-rolled segment of the steel business.

The plant is organized by the Aerospace Machinists. It has a wage rate considerably below that which the integrated steel companies pay.

Roanoke

Roanoke Electric Steel Corporation was one of the original minimills and a pioneer in many respects. The company was established in 1955, and, in 1962, installed one of the first continuous casters in the United States. Since that time, it has grown and replaced most of the facilities that were in operation at its inception. In 1983, an expansion program was undertaken which included the installation of a 90-ton furnace and a 5-strand continuous-casting machine. By 1986, it had four electric furnaces, two 25-ton furnaces plus one 40-ton furnace which was installed in 1986. Currently, its capacity on a 20-turn basis is 450,000 tons. With all furnaces operating, capacity can be as high as 500,000 tons. However, it is rare that the two small furnaces are in operation along with the two larger ones.

Roanoke began as a producer of reinforcing bar. In the 1960s, it started to produce other products such as angles and channels. In the 1970s, these new products constituted almost the entire output of the plant, since rebar production fell to approximately 2 percent of the total. There has been a resurgence since the demand on Roanoke for rebar increased as Bethlehem Steel dropped out of the market. It is now 10 percent of Roanoke's output.

Scrap is provided by the David Joseph Company, so that Roanoke is not required to shop around among scrap dealers for its needs. In addition, Roanoke operates its own shredder.

Productivity at the mill is somewhat lower by comparison with other minimills. It stands at about 3 1/2 man hours per ton. This is due to the fact that a wide variety of products are produced, requiring frequent roll changes. For example, several sizes and thicknesses of steel angles, as well as channels, are made. Other products are smooth bars and narrow flats.

As was the case with a number of minimills, Roanoke's first rolling facility was a cross-country mill. This has been changed and is now an in-line mill with 17 stands. This facility was installed in two phases, one in 1979 and the other in 1983. The current investment program involves a new reheat furnace (which cuts fuel consumption in half) and a replacement of the straightening and cooling bed. The entire project amounts to some $15 million and will be completed at the end of 1987. In 1985, the company considered installing a rolling mill to produce medium sections at a cost of $42 million. However, after further study, the project was abandoned. In addition to the steel mill, the company operates three bar-joist plants.

Roanoke's plant is nonunion. Although its set employment costs are somewhat lower than those of the integrated mills, it has a profit-sharing plan which permits sharing up to 20 percent of gross profits.

In terms of the future, the company will stay with its present product mix. It has no interest in entering the light flat-rolled segment of the steel industry.

Seattle Steel

Seattle Steel, Inc. is a newly organized company which purchased a plant in Seattle from Bethlehem Steel in 1985 for $45 million. Bethlehem's operation of the plant went back to the 1930s, when it was a cold-metal, open-hearth shop. Subsequently, electric furnaces were installed. Currently, there are 2, each with 120-ton heat capacity, giving the plant a 500,000-ton annual capacity to produce raw steel. The furnaces have water-cooled side panels and roofs.

After the plant was purchased from Bethlehem, the new company installed a continuous caster. Future plans call for improving the rolling facilities, which produce a large variety of products including rebar, 30-inch–wide plates, billets, structural sections, and tieplates for the railroads. Seattle is one of three plants in the United States producing tieplates, the other two being Bethlehem Steel at Steelton, Pennsylvania, and Razorback Steel in Arkansas.

The plant is organized by the United Steelworkers of America. However, the base rate is considerably below that paid by the integrated plants.

The company feels that its wide variety of products will allow it to operate successfully. It has no interest in entering the light, flat-rolled segment of the industry.

Sheffield Steel

Sheffield Steel Corporation is located in Sands Springs, Oklahoma. It was formerly a division of Armco Steel Corporation. The plant had a good performance record, however, in 1981, Armco decided to sell both Sheffield Steel and its other minimill plant, which is now Marion Steel, as a package. This hope was not realized so the plants were sold individually to two different groups.

Sheffield has two 85-ton furnaces with a capacity to produce 550,000 tons of raw steel. In addition, there is a continuous caster and a bar mill. Its production consists of rebar, fence posts, and billets.

Recently, the company purchased the Joliet, Illinois, plant of Continental Steel to which it ships 100,000 tons of billets annually to be rolled into merchant bars. Billets are also sold on a spot basis, some of which are sent to Armco. The scrap supply for the furnaces is drawn from Oklahoma and Kansas and is adequate to meet the company's needs. The furnaces operate efficiently with a tap-to-tap time of under two hours and man hours per ton for rebar production are less than two. Man hours per ton for billets, which are shipped in large tonnages, are less than one.

The company is organized by the United Steelworkers of America. However, the contract calls for a lower rate than that incurred by the integrated producers. In addition to the basic wage, the company has profit sharing as well as an incentive plan based on cost efficiency.

Sheffield feels that it has found its niche in the steel business and is not interested in entering the light, flat-rolled segment of the steel market.

Steel of West Virginia, Inc.

The company was founded in 1907 under the name of West Virginia Steel and Manufacturing Company. It was a rail reroller, located at Huntington, West Virginia, with no steelmaking capacity. In 1956, it was acquired by H. K. Porter and was made a division of Connors Steel. At that time, a 30-ton electric-arc furnace was installed. By the middle 1960s, a second 30-ton furnace was added, increasing the capacity from 84,000 tons to 150,000. It operated under this ownership until June 30, 1982, when it was shut down. Shortly afterward, it was purchased by the management and restarted in August 1982.

In January 1987, the Charter House Corporation bought 70 percent of the company, leaving the management with 30 percent of the ownership. At the present time, it operates two 70-ton furnaces with a capacity in excess of 300,000 tons. The plant does not roll the typical minimill products; rather, it produces light rails, small beams, and special sections. In addition to the electric furnaces, other facilities include a continuous caster and a blooming and bar mill.

The United Steelworkers of America have a contract with the company which calls for less than the employment costs of the integrated mills. Employees' incomes are supplemented by a bonus plan.

Conversations with the management indicate that there is no interest in entering the light, flat-rolled segment of the steel business.

Tamco

Tamco is located at Etiwanda in southern California, where it was founded as Ameron. The original plant had a capacity of 150,000 tons, with 3 relatively small electric furnaces, a continuous caster, and a rolling mill capable of producing bars as well as rods. In the late 1970s, the 3 furnaces were replaced by a 125-ton furnace, with an annual capacity of 175,000 tons.

In 1983, the ownership changed, as 50 percent of the company was acquired by two Japanese companies, Tokyo Steel and Mitsui Trading. Ameron, which still produces wire in the vicinity of the plant, owns the other 50 percent of the company. Currently, the plant has a capacity to produce approximately 300,000 tons. Products are concrete reinforcing bar and some rods that are sold to the construction trade for reinforcing concrete.

The original intention of the divided ownership with the Japanese companies was to export billets to Tokyo. This plan was later dropped.

The plant is unionized by the United Steelworkers of America, but with a lower employment cost than that of the integrated mills.

There is no interest in entering the light, flat-rolled-product sector of the steel industry.

Tennessee Forging

Tennessee Forging was established in the 1960s at Harriman, Tennessee. It subsequently acquired a plant at Newport, Arkansas. This second unit was sold and Tennessee Forging continued to operate at Harriman until 1983, when it was shut down. In 1986, the plant was bought by David Smith Associates and is expected to return to operations in mid- to late 1987.

The plant has electric furnaces with a capacity of 200,000 tons, continuous-casting equipment, and a cross-country bar mill. It will operate on a nonunion basis and produce minimill products.

The company has no intention of entering the light, flat-rolled segment of the steel industry.

Thomas Steel

Thomas Steel Corporation is located in Lemont, Illinois, near Chicago. The plant, which was sold to the current management in 1983, previously belonged to Ceco. Actual operations under the new mangement began in 1984. There are three 50-ton electric furnaces with water-cooled side panels, giving the company a total annual capacity of approximately 270,000 tons. In addition, there is a 3-strand caster and a cross-country rolling mill which produces a range of minimill products, including rebars, angles, flats, rounds, and squares.

The plant is not unionized. However, in addition to the base pay, there is a profit-sharing plan.

Despite its location near Chicago, steel is shipped as far as Pittsburgh. The company plans to continue to produce minimill products and has no interest in entering the light, flat-rolled segment of the industry.

Former Minimills

There are four companies that have at one time or another been classified as minimills. However, because of their current product mix, they do not fit into this category. They are Gilmore Steel Corporation, Marathon LeTourneau Company, Texas Steel Company, and Dibert, Bancroft & Ross Company, Ltd.

Gilmore Steel. Oregon Steel, a division of Gilmore Steel Corporation, was established in Portland, Oregon, to produce plates, which are not a typical minimill product. Although it is often referred to as a minimill, this classification is questionable because of its product. The plant was unusual insofar as it used iron ore, pumped ashore in slurry form, to be directly reduced in a Midrex facility. The resultant pellets were charged into an electric furnace. The steel was not continuously cast but pressure poured and then rolled into plates on a 108-inch, 4-high reversing mill.

In 1984, when the price of gas rose sharply, Gilmore closed down the direct-reduction facility. It now operates on 100 percent scrap.

Recently, the company reduced itself from two electric furnaces to one. With the application of new technology, such as eccentric bottom tapping and water-cooled panels, as well as the use of oxygen, one furnace has been upgraded so that it produces more than the two did previously. The revamped furnace produces 56 tons an hour whereas previously two furnaces together produced 43 tons an hour. The second furnace has been moth-balled. Total production is 370,000 tons, all of which is used for the production of plates.

The company is an ESOP with all 456 employees on salary. It has no intention of entering the light, flat-rolled sector of the steel industry.

Marathon LeTourneau. Marathon LeTourneau Co. is located in Longview, Texas. It has two electric furnaces, vacuum degassing, and rolling mills which make special-quality, alloy plates, and ingots for forging shops. Its capacity, when business warrants it, is approximately 100,000 tons a year. It does not have a continuous caster and uses a top as well as a bottom pouring process for its ingots.

Texas Steel. Texas Steel Co. was a minimill until the early 1980s, when it had 150,000 tons of capacity to make typical minimill products. Due to fierce competition, the company dropped out of the steel business and now functions as a steel foundry, making castings. The foundry capacity is approximately 18,000 tons a year.

Dibert, Bancroft & Ross. Dibert, Bancroft & Ross Co., Ltd., located at Amite, Louisiana, until recently was a minimill producing rebar and light structural sections. In the early 1980s, it ceased to melt steel for minimill products and has confined itself to a foundry operation. In late 1987, the company expects to resume rolling minimill products; however, it will not melt steel, but will purchase billets. The output will be in the 100,000-ton range.

Closures

Since 1975, closures and sales of minimills have included at least twenty plants. Some of them have been sold several times. For example, Pollack Steel Company, located near Cincinnati, was sold by the owners to Armco, which operated it for fifteen years and then sold it to a group of investors who have renamed it the Marion Steel Company, because of its location in Marion, Ohio. The Southern United Steel Company became the Southern Electric Steel division of Ceco Corporation, located in Birmingham, Alabama. It was subsequently sold to Birmingham Bolt, which in turn sold it to Birmingham Steel. Penn-Dixie Corporation, originally known as Continental Steel Company, was sold to Penn-Dixie and was closed in 1980, but reorganized as Continental Steel Company; it finally filed for liquidation in 1987.

Hurricane Steel Company, in Sealy, Texas, originally known as Shindler Brothers Company, was closed and Hurricane is now attempting to reopen it. Soule Steel Company, in Long Beach, California, was closed in 1985, as was Marathon Steel Company, located in Arizona. Youngstown Steel, organized in 1980 at Youngstown, Ohio, was closed down in 1983. Kentucky Electric, which operated for a number of years starting in the 1960s, went through a series of vicissitudes and was closed in 1985, only to be revived in 1986 as a division of Newport. Hunt Steel, organized in 1981, attempted to build a seamless-pipe mill, which did not work. The company went bankrupt and was closed and, subsequently, bought by North Star for $22 million in 1984. Southwest Steel Rolling Mills, at Los Angeles, California, was closed in 1977. The John A. Roebling Steel Company in Trenton, New Jersey, was closed in the late 1970s, then purchased and, in turn, sold to another group which finally closed it.

It is evident from these examples that not all minimills have been successful, and although a number of them have been closed down, they have reopened, since it was possible to purchase bankrupt companies for relatively little money.

Some of the plants seem to be permanently closed, such as Marathon in Arizona and Soule in southern California, but, like some integrated mills, these minimills have a habit of being reborn. Those that have been closed usually did not have modern equipment or found it difficult to compete for a number of reasons, including the high cost of power and vigorous competition from other minimills that have entered their territory.

Without question, the minimill operation will continue and, in many instances, will thrive. However, the more successful companies have had downstream operations.

Summary

From the foregoing analyses of the various minimill companies, a number of items stand out:

1. The minimills dominate the production of certain steel products, including concrete reinforcing bars, smooth merchant bars, and small structural shapes such as angles and channels, as well as narrow flats and small square sections. The integrated mills have, for the most part, moved out of these products, leaving them to the minimills.
2. The capacity has increased significantly as a number of mills have come into existence during the past two decades. In addition, the size of many of the mills has increased notably from the initial tonnage. The aggregate capacity has grown from approximately 3 million tons in the early 1960s to 19 million tons in 1987. This represents 17 percent of a total steel industry capacity of 112 million tons.
3. There have been many changes in ownership, in some instances as many as two or three, indicating financial difficulties.
4. The minimills have lower fixed employment costs than the integrated plants. However, in a number of instances, wages are supplemented by incentive plans and profitsharing.
5. There is adequate-to-surplus capacity for the type of product generally produced by the minimills. Therefore, expansion in this area will be minimal.

6. The trend toward multiple mills under one management has grown rapidly in the last few years and will probably continue, although at a somewhat decelerated rate.
7. The technology in the minimills has improved significantly from the rather crude installations in the early to middle 1960s to the more sophisticated equipment in the 1980s. This is particularly evident in the mills constructed and reconstructed in the last 10 years.
8. Very few minimills are interested in expanding into product lines such as light, flat-rolled products and seamless pipe and tubes, although two companies are currently involved in constructing plants to produce these products.

In regard to the ownership of multiple mills by one company, this has been achieved in two ways: 1) by constructing a number of new mills, which was the practice of Florida Steel and Nucor, and 2) by absorbing existing mills, which was the practice of Birmingham Steel and the predominant practice of North Star. It should be noted, however, that Florida Steel did acquire a mill in addition to those constructed and that North Star constructed two mills in addition to those acquired. Birmingham Steel has followed the acquisition route completely.

Other minimill companies that at present operate one plant have expressed an interest in acquiring additional facilities. These are usually financially stronger companies. It is expected that some of them will acquire the weaker companies in the not too distant future and thus expand their operations without adding new capacity to this segment of the industry, particularly since minimill facilities are in excess of the current and foreseeable demand for their product. Thus, the future tendency in the minimill sector will be to build balance sheets rather than steel plants.

In terms of the future, the possibility of minimills entering the light, flat-rolled or seamless pipe segments of the steel industry is extremely limited. Of the 29 companies interviewed, only five have an interest, while 24 have no intention, certainly at the present time, of making such a move. Nucor is already committed and has a light, flat-rolled plant under construction. North Star has constructed a seamless pipe mill, which will be put into operation at the end of 1987. North Star has also expressed an interest in entering the light, flat-rolled segment of the steel business provided Nucor's venture is successful. Other companies that have expressed an interest, although they have no plans, are Structural Metals and Chaparral. Georgetown is interested but by no means committed. This company, which uses 20 percent of direct-reduced iron in its electric-furnace charge, produces a steel of such quality that it could be more readily turned into sheets than that produced with a 100 percent scrap charge.

The technology for producing thin slabs is available. However, the capital expenditure required for the installation of this technology, along with the necessary rolling mill equipment, is much too large for the vast majority of minimill companies. In this respect, Birmingham Steel was approached by a foreign steel producer to form a light, flat-rolled products venture at a cost of some $200 million. The proposition was turned down, however. The company has subsequently contemplated the possibility of a joint venture.

Further, as will be indicated in this work, the demand for light, flat-rolled products, although great, is by no means equal to the potential supply available from the integrated producers. Any company entering this segment of the industry will find the market has not expanded. Therefore, it will have to compete with established facilities to carve out a share of a relatively stable market.

Future possibilities for the development of the minimill sector of the steel industry include the installation of downstream operations to use their products. Examples of this are Nucor, now the largest manufacturer of joists for the construction industry, and Florida Steel, with a number of plants for fabricating rebar. This downstream activity offers the greatest possibility for expansion, since the entrance into new product fields and the acquisition of other minimills is limited.

In addition to the integrated mills and the minimills, there are two other groups that fit neither category. The first is the specialty-steel producers, which operate electric furnaces but produce alloy or stainless steel and, in some instances, special-quality carbon steel. There are a large number of these, most of which are relatively small operations. Some of the principal large operations are Allegheny-Ludlum, Carpenter Technology, Copperweld, a division of Cyclops, Timken, Eastern Stainless, Quanex, Jessop Steel, and

Washington Steel. These companies were not included in the study, since they fit neither of the two categories for which a comparison was made.

The second group is constituted of those companies that are electric-furnace based and either too large to be considered minimills or too specialized in terms of product. Among these are Northwestern Steel & Wire; Lukens; Laclede; Colorado Fuel and Iron (as presently constituted); and Phoenix Steel.

A few of these companies have been classified by some as minimills. However, because of their size or their product, they are not considered as such in this book. Admittedly, this involves a judgment, but one must make a decision.

Notes

1. Florida Steel Corporation, *Annual Report,* 1986, p. 3.
2. Nucor Corporation, *Annual Reports,* 1983, 1984, 1985, p. 3.
3. Birmingham Steel Corporation, *Investor Presentation,* May 1986.
4. Birmingham Steel Corporation, *Annual Report,* 1986, p. 7.
5. Birmingham Steel Corporation, *op. cit.,* p. 5.
6. Ibid.
7. Birmingham Steel Corporation, *op. cit.,* p. 5.

4
Comparison of Minimills and Integrated Mills

The minimills and the integrated mills are part of the same industry, insofar as they both produce and sell steel products. There are, however, a number of significant differences between the two groups. These include size, type of products, geographic location, investment requirements, raw materials, productivity, technology, entry into the business, labor relations, employment costs, the impact of imports, and difficulties with closures and resale.

Size

The most obvious difference between the two groups is size. The minimills range from 150,000 tons of electric-furnace steelmaking capacity for an individual plant to well over 1 million tons, while the integrated plants range in capacity from about 1 million tons to as high as 6 to 7 million tons. A good average for the minimill is about 300,000 tons, while the average for the integrated mill is between 3 and 4 million tons.

The disparity in size when minimills first appeared in the early to mid-1960s was far greater. At that time, most of them were less than 150,000 tons, while the integrated mills in the early 1960s ranged from 1.5 million to as high as 8 to 9 million tons. Since then, there has been a decided growth in the size of the minimill and a shrinkage in the size of many integrated mills. In one sense, the smaller mills have been getting bigger, while the larger ones have become smaller. Outstanding examples of this are Chaparral, which rose from 400,000 tons to 1.5 million tons, and Inland Steel, which shrunk from 9.1 million tons to 6.5 million.

There has been some speculation that in the not too distant future, minimills will increase further in size and integrated plants will continue to shrink, so that the difference may be substantially narrowed. Further, some integrated plants have abandoned the coke-oven/blast-furnace/basic-oxygen complex as a source of steel production and have substituted electric furnaces, resulting in the use of the term *minimill-like plants.* Despite this trend, however, it is doubtful that the gap in size between the two will be reduced to a small amount. As long as one plant remains integrated and the other an electric-furnace operation, there will be a sizeable difference between the two. Based on average capacities of the two types of mills, the proportion is approximately 10 to 1 in favor of the integrated plant. Of their very essence, most electrics are small compared to the blast furnaces and BOF complex.

It is a fact that some minimill companies, such as Florida Steel, Nucor, and North Star, each have an aggregate capacity, when all of their plants are considered, that is in excess of some of the smaller integrated mills, such as Lone Star, Sharon, Acme, and McLouth. However, on an individual plant basis, there is only one minimill, namely Chaparral in Texas, that has a capacity comparable to the smallest integrated plants'.

Geographic Location

The locations of integrated steel plants and minimills were determined by different forces. Most of the integrated mills have been in operation at their present locations for a period of 80 to 100 years, whereas the majority of small electric-furnace plants have come into operation during the past 25 years.

Integrated Mill Location

Integrated mills were located in relation to raw materials as well as the market. Obviously, in a number of instances, there had to be compromises. The raw materials are basically iron ore and coal, which had to be converted into coke for use with the iron ore in the blast furnace. The locations of the two materials were limited. Coking coal came from western Pennsylvania, Virginia, West Virginia,

Kentucky, and Alabama, with some amounts available in southern Illinois. Ore, after 1900, came principally from the upper Michigan peninsula and Minnesota, although there were a number of deposits in other areas, such as Alabama, Pennsylvania, and New York.

By contrast, the raw material for the small electric-furnace mills, scrap, could be found throughout the country, and where it had to be transported, the distances were relatively short. Thus, the fundamental consideration for the location of minimills was the market. In fact, many have referred to minimills in recent years as "market mills."

At the turn of the century, the Pittsburgh area was recognized as the steel center of the United States. A number of plants built there in those days functioned until very recently. Some are still in operation. The location was considered ideal, since there was excellent coking coal and some ore in the area as well as river transportation for raw materials and finished steel products. However, shortly before the turn of the century, ore was brought to Pittsburgh in large tonnages from northern Michigan and Minnesota. Much of the journey was made by water through the Great Lakes to Conneaut on Lake Erie and then by rail to Pittsburgh.

Shortly after the turn of the century, the concentration of steel in the Chicago area built up rapidly. The most significant plant constructed in that area was the Gary works of United States Steel, built along with the town of Gary, Indiana, 30 miles east of Chicago, in the first decade of the century. Other companies, such as Inland, Republic, and the Steel and Tube Company of America (later a part of Youngstown Sheet and Tube), established themselves in this area, principally because it was developing into the largest steel market in the United States. Further, iron ore was readily accessible by water from Minnesota and the upper Michigan peninsula. One drawback was the need to haul coal by rail from western Pennsylvania, West Virginia, Virginia, and Kentucky.

At the present time, the Chicago area is still regarded as a place in which to be located. Bethlehem Steel tried to enter this area, once in 1930 and again in 1956, with two unsuccessful attempts to merge with Youngstown Sheet and Tube. Finally, in 1962, Bethlehem moved into the Chicago area with the construction of a new grassroots plant at Burns Harbor, somewhat east of Chicago. The plant in its present state was completed by 1970 and has a capacity for 5.3 million tons of raw steel.

The concentration in the Pittsburgh area has dwindled noticeably, as many of the plants in that region have been closed down in the 1980s. Pittsburgh became less desirable as the market grew in other areas, so that most of the investments of companies with multiplant locations were made in locations other than Pittsburgh. Consequently, the facilities were less efficient than those in other areas, so when capacity had to be reduced, the Pittsburgh plants were a prime target. In the 1960s, there were 25 blast furnaces in the Pittsburgh district. In June 1987, there was only one furnace operating and three that were operable.

Youngstown was considered a steel center until the 1970s. Plants had been built there shortly before the turn of the century, as well as during its first decade. Youngstown Sheet and Tube Company was headquartered there until it was absorbed by Jones & Laughlin in 1978. Other companies with integrated plants in the Youngstown area included United States Steel, Republic, and Sharon. This region has declined drastically as a location for integrated plants. United States Steel closed its plants and demolished the blast furnaces in the 1970s. Likewise, in the late 1970s and early 1980s Youngstown Sheet and Tube's two plants, Brier Hill and the Campbell works, were both closed permanently, except for the seamless-pipe mill which has been indefinitely mothballed. Republic, now a part of LTV Steel, has closed its plants there. Sharon, which recently filed for chapter 11 under the bankruptcy laws, has one operating blast furnace.

From as far back as the early 1960s, very little investment was made in the Youngstown area. When asked what equipment had been installed in his company's Youngstown plant, the chief executive of one of the companies with plants located there and elsewhere replied, "We have put a shear in." This was the sum of what he could recall.

An area that has had a minor degree of concentration is Birmingham, Alabama, and environs. United States Steel has a major facility there, as did Republic at Gadsden, Alabama, some 60 miles away. There were also a number of iron companies including Woodward, which had blast furnaces in the region, and James Walter, which still does. The attraction of the Birmingham area was the location of two mountain ranges close to each other, one with iron ore and the other with coal. The main plant of United States Steel is located between the two ranges.

Another location where several plants were constructed is the Cleveland area, which was and is the center of the iron ore trade, since the major independent producers, Cleveland Cliffs, M.A. Hanna, and Ogilvy-Norton, are headquartered there. Steel mills in the area include the Lorain works of United States Steel and the Cleveland plant of LTV Steel, which encompasses the plants formerly run by Republic and Jones & Laughlin which were adjacent to each other. The steel plants located in this region date back some 70 to 90 years.

Detroit has a concentration of three companies, which installed plants there to serve the automobile industry. These include Rouge Steel Corporation (a subsidiary of Ford Motor Company), Great Lakes Steel (one of National's plants), and Mclouth Steel (a relative newcomer to the steel industry, having been established in the early 1950s as an integrated plant). The advantages of the location are obvious. However, in spite of this, McLouth has been in financial difficulties for a number of years.

Other locations cannot be considered concentrations. On the East Coast, for example, there is the Sparrows Point plant of Bethlehem and the Fairless plant of United States Steel, about 100 miles apart. There was one integrated plant located in the western part of the United States before World War II, the Pueblo, Colorado, plant of Colorado Fuel and Iron Company.

There have been very few new locations for integrated plants established since 1920. Bethlehem Steel's plant at Burns Harbor, United States Steel's plant at Fairless Hills near Trenton, New Jersey, and the McLouth plant near Detroit are the only three new locations established since the early 1920s under normal conditions. There were four other new locations chosen during World War II by the Defense Plant Corporation. They included Fontana, California, for the Kaiser plant plus Geneva, Utah, which was considered out of range of Japanese bombers. Both were to furnish plates for the shipbuilding industry on the West Coast. In addition, there were two plants in Texas, one at Dangerfield (a blast furnace expanded into an integrated plant after World War II) and the other at Houston, a fully integrated unit which became part of Armco.

There are reasons for the continued operation of integrated steel plants at their original locations, almost all of which were established before 1920. This is particularly true of the Chicago area, which has remained an excellent location, since the market continued to grow in that area. In some other areas, where markets have declined, plants were maintained as long as the general U.S. market was strong enough to absorb their production, even though it had to be shipped by rail over considerable distances. When the market virtually collapsed in 1982 and did not recover in subsequent years, the poorly located plants were closed down. Some critics point to the locations as unfortunate choices, but this is hindsight. When the plants were built at their present locations, the market was there and the raw materials were available. This showed good judgment at the time, and in some instances, it is still vindicated as good judgment.

Because of the large capital investment required for a new integrated steel plant, many of the plants have remained at their original locations, even though raw materials as well as finished products have to be shipped over long distances. It is interesting to note that there have been very few plants built at new locations in the past 70 years. A review of the current integrated plants indicates this condition.

United States Steel:

The Gary, Indiana, plant was built from 1906 to 1908 and still is at the original location.

Lorain, Ohio works was acquired when United States Steel was formed in 1901 and is still at the original location.

Fairfield works (formerly part of Tennessee Coal, Iron and Railroad Company) was acquired by United States Steel in 1907 and is still at the original location.

Edgar Thomson works in Pittsburgh was acquired when United States Steel was formed in 1901 and is still at the original location.

Fairless works near Trenton, New Jersey, was built by United States Steel from 1950 to 1952. It is one of the few new locations for an integrated plant in the past 60 years.

Geneva works at Geneva, Utah, was acquired by United States Steel after World War II. It was built in this location during World War II in order to be safe from possible Japanese bombing raids.

The plant has been part of United States Steel for 40 years, but has been permanently closed, although there is some activity underway to reopen it.

Bethlehem Steel:

Bethlehem, Pennsylvania was founded at that location in 1860 and maintained there since.

Sparrows Point near Baltimore, Maryland, was built in the 1880s by the Maryland Steel Company, acquired by Bethlehem in 1916, and located at the same place since then.

Lackawanna near Buffalo, New York, was built by the Lackawanna Steel Company in 1902, acquired by Bethlehem in 1922, and operated at that location until it was shut down in 1984.

Burns Harbor, Indiana, a fully integrated plant, was built by Bethlehem in the 1960s. This is the latest integrated plant built in the United States and represents a new location.

Johnstown and Steelton, Pennsylvania were both integrated, but have been transformed into electric-furnace operations. The Johnstown plant was acquired in 1922 from the Cambria Steel Company. It had been at that location since the mid-1850s. The Steelton plant, acquired in 1916 from the Pennsylvania Steel Company, was built at that location in 1867 by the Pennsylvania Railroad to provide rails.

LTV Steel:

Indiana Harbor plant was built by the Steel and Tube Company of America in the second decade of the century and acquired by Youngstown Sheet & Tube in the 1920s. It was taken over by Jones & Laughlin in 1978. It is one of the key plants in the merger of Republic and Jones & Laughlin to form LTV Steel.

Cleveland plant results from the merger of the Jones & Laughlin and Republic facilities located next to each other. Both plants were built before 1920.

Warren plant in Warren, Ohio, was built before 1920 by the Trumbull Steel Company and subsequently acquired by Republic Steel.

Gulf States plant in Gadsden, Alabama, owned originally by the Gulf States Steel Company, was built before 1920 and acquired by Republic in the mid-1930s. It is now functioning as Gulf States Steel under independent management.

Armco: Has two integrated plants, one in Ashland, Kentucky, and the other in Middletown, Ohio. The Ashland plant was acquired in 1920 from the Ashland Iron and Mining Company, while the Middletown plant, which was the original facility of the company, was integrated insofar as iron was supplied from blast furnaces located at Hamilton, some 12 miles away. In 1953, a blast furnace was constructed at the plant in Middletown. The Ashland plant was constructed at its present location in 1869, with its furnaces rebuilt in the early twentieth century and again in the post–World War II period. Thus, there has been iron and steel activity at this location for over a century.

Inland Steel: The Indiana Harbor plant was established at its present location in 1901.

National Steel: Currently operates two integrated plants, one near Detroit (built in 1923 by the Michigan Steel Corporation) and the other at Granite City, Illinois, established at that location in the last century.

Weirton Steel: Is a newly established company as the result of the plant's purchase from National Steel by the employees in 1984. Its plant was built at the present location in 1918.

Rouge Steel: A wholly owned subsidiary of Ford Motor Company, was established at its present location in 1919.

Wheeling-Pittsburgh: Is a result of a merger between Wheeling Steel and Pittsburgh Steel in 1968. It operated two integrated plants, one at Monessen, Pennsylvania (built in 1912 and recently abandoned) and the other, which is still functioning in the Wheeling-Steubenville area. The Wheeling plant dates back to 1874.

McLouth Steel: Its plant at Trenton, Michigan, near Detroit, represents one of the few new sites established for an integrated steel plant since 1920. The plant became integrated in 1953.

Lone Star: It is an integrated plant, although its blast furnaces and open hearths have been shut down since 1986. The plant, located at Dangerfield, Texas, is a World War II product and has functioned since that time. It was at first a blast furnace, but steelmaking and finishing facilities were added in the postwar period.

Although the above locations cited are the same, the facilities have been replaced and updated several times.

Minimill Location

In contrast to the long-lived existence of integrated plants in their original locations, the minimills, springing up as they did in the 1960s and 1970s, were established in areas where there was a growing market for their products, and this was usually away from the integrated mills. As previously indicated, there was little need to be concerned about raw materials, since they all depend on scrap, which is usually available in sufficient quantities in almost any populated section of the country. It should be noted that, like the integrated mills, the minimills, once established, have remained at their original locations.

In analyzing the minimill sector of the industry, one finds that there are 20 minimills in the southeastern quadrant of the United States. Most of these were built in the 1960s and 1970s, although some predate this period. The mills in the southeastern United States are:

Florida Steel: Tampa, Florida; Jacksonville, Florida; Jackson, Tennessee; Knoxville, Tennessee; and Charlotte, North Carolina

Nucor: Darlington, South Carolina

Birmingham Steel: Birmingham, Alabama; Jackson, Mississippi; and Chesapeake, Virginia

Structural Metals: Birmingham, Alabama

Newport Steel: Newport, Kentucky; and Ashland, Kentucky

Bayou Steel: LaPlace, Louisiana

Owens Steel: Columbia, South Carolina

Roanoke Steel: Roanoke, Virginia

Georgetown Steel: Georgetown, South Carolina

Atlantic Steel: Atlanta, Georgia; and Cartersville, Georgia

Tennessee Forging Steel: Harriman, Tennessee

Steel of West Virginia: Huntington, West Virginia

Some of these plants have been at their present location for many years. For example, the location at Huntington, West Virginia, was in use as far back as 1907, when the West Virginia Steel and Manufacturing Company was established. The Newport, Kentucky, location was established in 1909. The Atlantic Steel location in Atlanta was established in 1901. Most of the other plants are of recent vintage, having been established in the last 25 years.

The attraction of the southeast location was its rapid growth in population between 1950 and 1980—from 27 million to over 44 million people. Such an increase brought with it additional demand for steel products and, in particular, those made by the minimills for housing and service industries, such as shopping centers. This type of construction used large quantities of reinforcing bar and small structural sections.

The Southeast's increase in population has leveled off somewhat in the 1980s, and the 20 minimills in the area now find their capacity in excess of demand, so much so that they are in keen competition with each other and are now reaching out beyond the area to market their products.

Outside of the southeastern quadrant of the United States, there is a concentration of minimills in Texas and Oklahoma, where seven are located, and also in the North Central states, in and around the Chicago market, where there are eight mills. The seven mills in the Texas/Oklahoma area include:

Chaparral Steel: Midlothian, Texas

Structural Metals: Seguin, Texas

North Star Texas: Beaumont, Texas

Border Steel: El Paso, Texas

Nucor: Jewett, Texas

Sheffield Steel: Sand Springs, Oklahoma

Hurricane Industries: Sealy, Texas

Mills located in the North Central states are:

Birmingham Steel: Kankakee, Illinois

Thomas Steel: Lamont, Illinois

Calumet Steel: Chicago Heights, Illinois

Charter Steel: Chicago, Illinois

Marion Steel: Marion, Ohio

North Star Steel: Youngstown, Ohio; Wilton, Iowa; and Monroe, Michigan

Other minimills are scattered, with some in the northeastern part of the United States and a few on the West Coast.

The type of products produced by the minimills are in demand wherever there is growth in population. Thus, a number of companies were encouraged to establish minimills in growth areas, and, in spite of the present market saturation, very few have closed down, although many have changed ownership—some more than once. The basic reason for the continued existence of these plants is the very small amount of investment required to purchase and revive a minimill. An example of this is the Tennessee Forging operation, which was closed down and revived recently, when it was purchased by David Smith Associates. Connors Steel is another example. The plant was shut down by H.K. Porter in 1983 and purchased and revived by Structural Metals within the year. There seems to be little doubt that most of these mills will continue in operation under their present or new ownership, when one considers that there are some 50 mills operating and fewer than 10 have been shut down permanently.

In terms of new construction, there is little hope that the number of traditional minimills will be expanded, since the market for their products is saturated. The investment required for a new minimill today is substantially above what it was in the 1960s and 1970s. The most recent plant additions were by Nucor and Florida Steel at a cost of $55–60 million. Further, there are few areas that would require the construction of a minimill to satisfy local demand. It should be noted that minimills are now reaching out to areas far beyond the original 200–300 mile radius that was considered normal in the 1960s and 1970s.

In the case of the integrated mills, where plants are being closed, there is no hope for expansion. The minimills also have reached a saturation point. It is possible that one or two additional units, producing traditional minimill products, may be installed during the next decade, but this is not likely.

Raw Materials

Minimills

The minimills throughout the United States, with one exception, depend on scrap for 100 percent of their raw material. The exception is Georgetown Steel Corporation's facility in South Carolina, which can produce 400,000 tons of direct-reduced iron a year.

Direct reduction is a process whereby deoxidized iron pellets are produced without the use of a blast furnace. Iron pellets with approximately 60 percent iron content are fed into a shaft furnace. Natural gas is also introduced and removes much of the oxygen from the iron ore so that the resultant pellet is approximately 90 percent iron. This is relatively free of contaminants and is then charged into the electric furnace. At Georgetown, this currently represents about 20 percent of the furnace charge with the remainder made up of scrap.

By contrast, the raw materials for the integrated plants are basically iron ore and coal, which must be converted to coke. Both of these are then charged into the blast furnace along with limestone to produce molten pig iron. The resultant molten pig iron is charged into a basic-oxygen converter along with scrap to produce steel; the ratio is generally 70–75 percent molten pig iron to 25–30 percent scrap. Thus, both the minimills and the integrated mills use scrap in order to make steel. However, the dependence of the integrated plant on scrap is much less than that of the minimill.

Scrap is generally available throughout the country with heavier concentrations in some areas as compared to others. Basically, it comes from three sources:

1. *Mill-generated scrap,* which is produced as raw steel, is cropped and sheared while passing from one process to another on the way to a finished product. This is recycled and accounts for a large portion of the scrap used by the integrated plants. Mill-generated scrap has declined in tonnage because of the installation of continuous casting, which improves the yield of finished product from raw steel. Prior to the adoption of continuous casting, yield was about 72 percent, with 28 percent of raw steel recycled as scrap. With the installation of continuous casting throughout the steel industry (including minimills and integrated mills), the yield has improved considerably and is now in the 80–85 percent range. In the minimill, where there is usually just one operation (the reduction of billet to a bar), the yield is much higher, running well over 90 percent, so that the amount of mill scrap available to the minimills is much less than that available to the integrated mills due to the fact that they have many more processes and generate more scrap. Mill-generated scrap is considered excellent scrap, particularly in the mills that concentrate on the production of carbon steel products.
2. *Prompt industrial scrap,* which is generated as steel users fabricate the steel into finished products, such as automobiles, refrigerators, and machinery. This is usually very good scrap, since its content is known.
3. *Obsolete metal,* consisting of discarded items that contain steel ranging from automobiles, to old machinery to multistory buildings that are torn down. This source provides a substantial amount of scrap, which must be processed before it can be used in the steelmaking furnaces. There are a number of grades of obsolete metal, ranging from no. 1 heavy-melting scrap down through no. 2 bundles. Based on the type of scrap, prices also cover a reasonably wide range. For example, in early 1987, the price of no. 1 heavy-melting scrap, as reported in the American

Metal Market, was $80 a ton in Pittsburgh, $75 in Chicago, $60–62 in Birmingham, Alabama, and $78 in Seattle. No. 2 bundles cost $47 in Pittsburgh, $59 in Chicago, and $47 in Birmingham.[1] Prices of scrap also fluctuate violently from year to year. For example, in 1981, the average price in the Pittsburgh district for no. 1 heavy-melting scrap was $100.57, while in 1985, the average was $77.33. Because of the fluctuations in price from year to year, as well as from place to place, minimill costs of production will vary considerably, reflecting this difference in price.[2]

Greater demands have been placed on electric furnaces to produce higher quality steel. As a consequence, it is necessary for the operators of these units to be more selective in the scrap they charge into the furnaces in order to keep the contaminants at a minimum. This has given rise to problems for some producers. It is a fact that any and all steel products can be made from electric-furnance steel, provided it is pure enough; to achieve this objective, the scrap must be virtually free of contaminants or have them diluted to a great extent with the addition of direct-reduced iron or pig iron.

To sum up, there is an abundance of scrap in the United States, particularly following the decreased steel production experienced in the past few years. The problem that exists is one of quality, not quantity. In order to insure their supply of scrap, a number of minimills have tied in with scrap dealers as exclusive suppliers. Florida Steel, for example, uses the David Joseph Company as a sole supplier.

Integrated Plants

The integrated plants require huge amounts of raw materials that are concentrated in a few locations, as opposed to scrap found in varying quantities throughout the country. The overwhelming percentage of the iron ore produced in the United States comes from the northern peninsula of Michigan and Minnesota. Much of the ore mined in this area is low-grade taconite with an iron content varying from 25 to 30 percent. This is extremely hard material and must be crushed before it can be concentrated and formed into pellets with an iron content of 64 to 65 percent. The facilities required for this processing are expensive. The most recent unit placed in operation, which is capable of producing 4 million tons a year, required an investment of approximately $90 per ton of capacity or $360 million. Ore is also found in other areas in the country, such as Alabama, Texas, and Pennsylvania. These are relatively small deposits compared to the Michigan/Minnesota reserves.

A significant amount of ore is imported. In 1979, a year in which 136 million tons of raw steel were produced, iron ore imports were 33.8 million tons, 22.6 million of which came from Canada. Venezuela, Brazil, and Liberia accounted for 4.5, 3.1, and 2.2 million tons, respectively. In recent years, with the amount of steel produced declining, ore imports also declined, reaching a low of 13.2 million tons in 1983 and recovering slightly to 16 million tons in 1985. Again, Canada provided more than half, with Venezuela, Brazil, and Liberia providing approximately 2 million tons each.

A number of integrated companies have interests in Canadian ore deposits. The Iron Ore Company of Canada and Wabush Mines, both located in the Quebec/Labrador area, have significant participation by U.S. integrated companies, including Bethlehem, National, Inland, and Acme, while United States Steel's interest is confined to the Quebec-Cartier Mining Company. Imports of Brazilian, Venezuelan, and Liberian ore are delivered predominantly to the East Coast and southern states, while the Canadian ore is in use in the plants in the Great Lakes area as well as the East Coast.

In addition to pellets, a considerable amount of sinter is charged into the blast furnace. This is produced by mixing fine iron ore with coke breeze or anthracite coal fines. The final product is a clinkerlike substance that is desirable in the blast-furnace operation. It should be noted that iron ore is no longer a raw material when it has been charged into the furnace. In the form of pellets or sinter it is definitely a manufactured product.

The other basic raw material, coal, which must be metallurgical coking coal, is found predominantly in western Pennsylvania, West Virginia, Kentucky, Alabama, and, to some extent, southern Illinois and Oklahoma. This coal is converted into coke in ovens located at or near the integrated steel mills.

Coke ovens represent a large investment of approximately $225 to $250 per ton of capacity. In a modern battery of ovens, capacity ranges from 700,000 tons to 1 million tons.

There is an abundance of coking coal in the United States. In fact, exports of this material are in excess of 60 million tons. Reserves run into the billions of tons. The problem here is not in the coal, but in the coke ovens—many of which are quite old and will need replacement. At the current rate of operations, the existing coke ovens are adequate. If demand for steel should increase significantly, there could be a problem.

Raw materials for the integrated plants are available in terms of both quantity and quality. This is not equally true of scrap for electric furnaces which, although available in quantity, presents some difficulties in terms of quality.

Technology

The technologies of integrated mills and minimills vary widely, yet in some respects they are similar. With few exceptions, the minimills have been built before 1980, and the facilities of these mills need some updating if they are to be modern and competitive. The integrated mills, with but three exceptions (the Fairless works of United States Steel, the Burns Harbor plant of Bethlehem Steel, and McLouth near Detroit), were all built before or during World War II. Many of their facilities have been replaced and modernized using, in most cases, the latest available technology. This is a continuous task as new technology is developed. The same norm applies to minimills.

The minimills operate electric furnaces. Here, the improvements include water-cooled side panels and roofs as well as high-powered transformers and the use of oxygen. A number of minimills have updated their electric furnaces to include these improvements. However, a number have not. Consequently, in the melting process in some of the smaller, older minimills, transformers are low-powered, and water-cooled panels and roofs have not yet been installed. A number of minimills, such as New Jersey Steel and Structural Metals in Texas, have recently replaced smaller furnaces with larger, modern ones. In terms of continuous casting, the minimills, virtually without exception, have continuous-casting units through which the steel is poured and formed into billets. Some of these units are older than others, but all function reasonably well. However, the older ones in many instances have higher costs.

As far as bar mills are concerned, many of these were cross-country mills with looping devices. In most cases, they have been replaced by straightaway mills with one stand after another in-line. These are more desirable than the cross-country mills, yet some plants still operate a combination of cross-country and in-line mills. However, even though a number of mills lack the most modern equipment, what they have is adequate for the production of minimill-type products, although not fully competitive.

The rod mills at Raritan River, North Star Texas, and Georgetown are state-of-the-art. Some of the bar mills, such as those installed in the recently constructed mills of Florida Steel, Bayou, and Nucor, are thoroughly efficient, as is the new mill installed at Cascade to replace the obsolete, older mill.

The proposed Nucor mill for the production of flat-rolled products represents the adoption of new technology. The slab to be cast is much thinner than any normally cast for the production of steel sheets. Further, the hot-strip mill has only four stands. These facilities represent a new approach to the production of steel sheets, and it remains to be seen how successful they will be. In the production of medium to heavy structural steel sections, both Chaparral (at its existing facility) and Nucor (at its proposed facility) will have the latest equipment.

Since most of the integrated mills have been in operation for decades, there is a constant need to upgrade the technology. This has been done in almost every phase of production. Particularly significant is the installation of continuous-casting machines, which allow the molten steel as it is poured from the ladle to be cast through this equipment and come out in the form of a slab, billet, or bloom. Virtually all of the major mills have installed these units during the past few years, and in this respect, they are technologically up to date.

In the preparation of materials for use in the blast furnace, significant advances have been made. Iron ore has been beneficiated and either sintered or formed into pellets for use in the blast furnace. This upgrades the amount of iron in the burden from what was about 50 percent iron content in raw iron ore to 62–65 percent iron content in the prepared pellet. With the improvement in furnace burden, the blast-furnace output has been increased significantly. Blast furnaces that produced 1,500 tons of iron a day in the middle 1950s, now are capable of almost twice that much. Further, the improved furnace practice has reduced the amount of coke needed to produce a ton of iron from almost one ton to a little over half a ton.

The steelmaking process has been changed from the open hearth to the basic-oxygen furnace in the past 25 years. This provides a more efficient and less costly operation. In the rolling mills and finishing mills, significant advances have been made. Perhaps, the most important advance has been in the continuous hot-strip mill. All of the sheet products, including hot-rolled, cold-rolled, galvanized, and various specialty sheets, must pass through the hot-strip mill, and this accounts for over 50 percent of all steel products. Beginning in the early 1960s and extending through the mid-1970s, some 12 new hot-strip mills were installed in various plants throughout the United States. They differ so much from the mills installed prior to that time that they are called the "second-generation" hot-strip mill.[3] In spite of their relatively recent installation, additional upgrading is necessary for them to be competitive on a worldwide basis. This is underway currently at a number of mills.

Other notable advances in technology include the seamless-pipe mills of United States Steel at Fairfield, Alabama, which embrace the latest technology; the two cold-reduction mills for the production of sheets, which will be installed by the joint venture of United States Steel and Pohang Steel in Pittsburg, California; and the joint venture of Inland Steel and Nippon Steel, which will be installed in Indiana. These will differ from the conventional four-high mill insofar as each of the stands will have six rolls to produce sheets that are almost perfect in terms of gauge. The investment required in Pittsburg and Indiana is approximately $400 million for each.

The capacity of the steel industry in the United States has been reduced from 160 million tons in 1977 to approximately 112 million tons in 1987. A very substantial proportion of the 112 million tons consists of modern integrated mills and minimills with competitive facilities which employ the latest technology.

Investments

The difference between the minimills and the integrated mills in size and complexity of facilities accounts for the vast difference in the amount of investment required for each. To attempt to compare the investments would be an exercise in futility, since the facilities of the integrated plants are so much larger and embrace so many more operations than those of the minimills. Further, the amount of steel produced and the fact that the products differ so widely make comparison virtually impossible.

Bethlehem Steel built the last integrated plant in the United States in the 1960s at Burns Harbor, near Chicago. Currently, the plant has a capacity to produce 5.3 million tons of raw steel and to transform them into a variety of sheet and plate products. The facilities include coke ovens, blast furnaces, basic-oxygen steelmaking converters, primary rolling mills, continuous casters, plate mills, a hot-strip mill, cold-reduction mill, and galvanizing equipment. The actual investment in the 1960s was somewhat in excess of $1.5 billion. The replacement value of the entire plant, in terms of today's costs, has been estimated at $4.6 billion.

By sharp contrast, one of the most recent installations in the minimill sector is the Jackson, Tennessee, plant of Florida Steel. This plant includes scrap-handling facilities, one electric furnace, a continuous caster, a bar mill, and a cooling bed. Total capacity is approximately 400,000 tons. In contrast to the millions of tons of plates and sheets—hot, cold, and galvanized—produced at Burns Harbor, the Jackson plant ships a little less than 400,000 tons of minimill products, such as concrete reinforcing bars, hot-rolled smooth bars, and light structural shapes. The investment involved was about $60 million.

The difference in capital investment is staggering. However, it would be deceiving to take the capital cost totals for Burns Harbor and Jackson to make a cost-per-ton comparison and conclude that minimills can install capacity for about $150 per ton versus $868 per ton for their integrated counterparts. In spite of the vast differences in product mix and tonnage, such calculations are often made in ascribing capital cost advantages to minimills as though they embody the same complexity and capabilities as the integrated mills. This is simply not the case.

In addition to the initial investment for both the integrated and minimills, other expenditures are made from time to time to replace and upgrade individual facilities in the plants. For example, the United States Steel plant at Gary, Indiana, recently acquired a continuous caster at a cost of $260 million. Further, the cost of relining and upgrading the blast furnace at the Indiana Harbor Works of LTV Steel is in the neighborhood of $117 million. The seamless-pipe mill installed at the Fairfield works of United States Steel, along with a continuous caster, required an investment of some $750 million.

Replacements at the minimills of course cost much less. At Florida Steel, the Tampa plant will be gradually replaced by a new electric furnace, a caster, and a rolling mill. The total expense will be in the area of $60 million.

From time to time, comparisons have been made on the basis of facility cost, indicating that the minimill requires so much less investment. However, if we compare the casting machine at the Gary works with that which Florida Steel will install at Tampa, the differences are dramatic. The caster at the Gary works is capable of handling slabs up to 80 inches wide and 10 inches thick, with a total capacity to produce 3.5 million tons, while that at the Tampa plant will cast billets approximately 4-inches-by-4-inches square, with a total capacity of some 300,000 tons.

Investments in the integrated segment of the steel industry for the next decade will be confined to improving and upgrading as well as replacing current facilities. There will be no new greenfield-site plant constructed. In regard to the minimills, there will be considerable investment in upgrading and improving existing facilities, with the possibility of one or two new standard minimills being constructed.

New Products

Mention has been made of the plants currently under construction by Nucor to produce sheets and large structural sections, as well as that by North Star for the production of seamless pipe. In relation to the sheet mill, it is very doubtful that other minimills will build facilities for sheet production, with the possible exception of North

Star and Georgetown Steel. These sheet mill plants involve a much larger capital outlay than the traditional minimills. As indicated, Nucor's flat-rolled products plant, which is far simpler than its counterpart in the integrated segment, will require an investment of $225 million. The structural mill, which is a joint venture, is projected to require $190 million.

The low cost of the sheet mill (compared to that of the integrated plant performing the same function) is due to the difference in the basic equipment of each. The integrated plant must invest heavily in coke ovens, blast furnaces, basic-oxygen steelmaking capacity, continuous casting, and a huge 80-inch–wide hot-strip mill, while Nucor will rely on electric furnaces, a caster to produce thin slabs, and a scaled down 54-inch–wide hot-strip mill, since the thin slabs will require less reduction. It should be kept in mind that the integrated plants have already made their investment in these facilities and, in the future, will require relatively small amounts of money to keep them in operation and improve them, while Nucor's investment is still ahead of it. Further, in spite of the lower investment cost, since Nucor's plant has new technology, the question arises as to how successful the operation will be. A further question arises as to whether or not more flat-rolled capacity is needed in the United States.

The investment at Nucor will provide hot-rolled and cold-rolled sheets, whose quality would not be acceptable for many applications. Further, it does not involve the production of tinplate, galvanized sheets, or electrical sheets, which integrated plants are able to produce because of their extensive investment in the appropriate facilities.

Productivity

Comparison have been made between the productivity figures of the minimills and those of the integrated mills in terms of man hours per ton shipped. For the minimill, the figures vary from less than 2 hours to 4 hours; a good average would be 2 man hours per ton. On the other hand, productivity figures for the integrated mills are greater, varying from 4 man hours per ton shipped to 7 or 8 man hours.

At first sight, this looks as if the minimills are more efficient than the integrated operations. However, one should examine the data fully before reaching a conclusion.

The minimills have three basic operations:

1. Scrap is melted in an electric furnace to produce liquid steel.
2. Liquid steel is poured through a continuous caster to produce semifinished steel, usually in the form of billets.
3. The billets are rolled hot on a bar mill to produce products, such as reinforcing bars, small structural sections, and rods, which are then allowed to cool.

The integrated plant has many more operations, starting with basic raw materials such as iron ore and coal. The iron ore has to be transformed into pellets or sinter, while the coal is turned into coke. Both of these items are charged into a blast furnace along with limestone to produce molten pig iron. All of these steps are not included in the minimill operations. When the iron is produced, it is charged in molten form along with scrap into a basic-oxygen furnace, where it becomes steel. The liquid steel is then poured through a continuous caster and emerges in the form of either a bloom, billet, or slab. These are then rolled hot into various products. In the case of the sheets, which constitute the major product of the integrated mills, slabs are hot-rolled on the continuous hot-strip mill and emerge as hot-rolled sheets. These can be sold as such or further processed into cold-rolled sheets. This requires a pickling operation which cleans the surface of the steel. After this the steel is cold-reduced and then annealed to improve its physical qualities. This steel can be sold as cold-rolled sheets. However, some of these are further processed with zinc coating to form galvanized steel, or they can be coated with tin to form tinplate. Thus, the integrated mill performs many more operations than the minimill, and this accounts for higher man hours per ton of product shipped.

If the comparison is to be valid, the minimill man hours should be compared with the man hours from comparable operations in the integrated mill. The operations to be compared would be steelmaking, continuous casting, and hot rolling. The operations that precede steelmaking should not be included, nor should those that follow the hot-rolling process, since the minimills do not have these functions. If a comparison is made between the man hours needed in the three steps in the integrated plant that correspond to those of the minimill, it has some validity.

As previously mentioned, minimill man hours per ton range from fewer than 2 to 4. By comparison, man hours per ton required for the corresponding operations in the integrated plant are somewhat less. Based on an examination of several integrated plants, the man hours per ton for the three functions (namely, steelmaking, continuous casting, and hot rolling) range from fewer than 1 to 2.5. In one of the plants, the figure amounted to approximately three-fourths of a man hour (actually, 0.758 man hours per ton). In another, it ranged from 1.1 to 1.2. In a third, it was 2.5. The fourth mill recorded 0.794 man hours for these operations.

Man hours per ton in the minimill vary according to the size of the product rolled. For example, at one minimill, when size 8 rebar (which is about an inch in diameter) is rolled, man hours total 1.1, and when size 3 (which is three-eighths of an inch in diameter) is rolled, man hours increase to 1.7. A number of integrated plants require fewer man hours to support comparable phases of minimill production largely because of the economies of scale, whereby the labor input is spread over a considerably greater output. In the steelmaking phase, for example, the electric-furnace shops operated by minimills can melt an average of 330,000 net tons per year, while the BOF shops in the integrated mill can produce an average output of 2.8 million net tons per year. Subsequent operations reveal comparable disparities due to plant scale. This is true of the continuous-casting and hot-rolling phases of production. The much larger tonnages produced by the integrated mills work to reduce the relative labor input required. In the minimills, positive changes in productivity often result from increases in throughput that come with changes in product mix, as indicated in the case cited previously. The three basic steps in the minimill operation, as well as the multiplicity of functions and processes in the integrated mill, are given graphically in the diagrams of each found in chapter 1.

Employment Costs

One of the interesting comparisons between the minimill and the integrated plant is employment costs. With very few exceptions, the integrated mills are organized by the United Steelworkers of

America. The most notable exceptions are the Middletown, Ohio, plant of Armco and the Weirton Steel plant, both of which are organized by independent unions. Throughout the United States, more than 50 percent of the minimill companies, representing more than 60 percent of the capacity, are nonunion. In the instances where minimills are organized, there are other unions in addition to the United Steelworkers. At the unionized minimill plant, employment costs are considerably below those of the integrated plants. While costs vary in the integrated mills, most plants have costs above $20 an hour. The minimills' basic employment costs, in contrast, are from $8 to $10 an hour less.

A number of minimills operate on a profit-sharing plan or an incentive system, so that while the basic hourly employment rate may be low, the employees are able to earn almost as much as the unionized steelworkers in integrated plants. The profit-sharing and incentive plans have a distinct advantage and make a significant contribution to productivity. The incentive plans in some plants call for pooling a certain number of dollars per ton over and above a stipulated tonnage, and this is divided among the workers in the unit. Very often, when one worker retires, the others petition management not to replace the person so there can be fewer workers to divide the pool. In one plant, the incentive standard is set at about 50 percent of the facility's capability, so that every ton above the 50 percent earns additional pay for the crew. Another advantage in terms of employment costs and productivity is the minimum amount of work rules in the minimill, which provides for an effective use of the labor force.

In the recent labor negotiations conducted between the United Steelworkers of America and the integrated steel producers, significant wage concessions were made, ranging from 40 cents an hour at Inland Steel up to $3.65 at LTV. Further, changes were made in the work rules, so that the work force could be employed more flexibly. This change will allow for an increase in productivity. On the part of the companies, profit-sharing plans have been agreed to, as well as concessions involving stock.

Steel Imports

Steel imports that compete with minimill products generally fall into two categories: 1) concrete reinforcing bars and 2) bar shapes under three inches. There are also some products in the category of structural shapes three inches and over. However, many of the structural items produced by minimills are under three inches.

In the past 25 years, imports in these two categories have fluctuated considerably. The most troublesome for minimills recently has been concrete reinforcing bars. They hit an all-time high in the 1960s, peaking in 1968, when 740,000 tons of rebar entered the country. That same year constituted a new high for total imports—18 million tons. In 1969 and 1970, the tonnage dropped sharply when voluntary restraint agreements with a number of countries were in force. These arrangements caused the exporting countries to shift to higher-value products.

After 1971, rebar imports never reached 500,000 tons. Their behavior has been erratic, dropping to 142,000 tons in 1975. During the past 10 years, they have fluctuated violently, as table 4–1 indicates. In the mid- to late 1970s, there was a dramatic drop as the tonnage averaged approximately 100,000 tons for the years 1977–79. It continued to fall to all-time lows of 52,600 and 51,700 tons in 1981

Table 4–1
Imports of Concrete Reinforcing Bar, 1965–86
(thousands of tons)

Year	*Imports*
1965	568
1966	673
1967	567
1968	740
1969	471
1970	212
1971	515
1972	358
1973	286
1974	478
1975	142
1976	192
1977	93
1978	110
1979	116
1980	79
1981	53
1982	52
1983	208
1984	434
1985	411
1986	448

Source: American Iron and Steel Industry, *Annual Statistical Reports* for selected years.

and 1982, respectively. There was a resurgence in 1984, 1985, and 1986, as the figure topped 400,000 tons in each of these years.

In the past few years, the source of rebar imports has changed significantly. In the late 1960s and early 1970s, the European Community provided a considerable portion of the total tonnage, as did Japan and Latin America. In 1968, Belgium and Luxembourg provided 285,000 tons of the 740,000-ton total. France added 95,000 tons; Germany, 82,000 tons; and Japan, 70,000 tons. In 1974, with a total of 478,000 tons, Belgium and Luxembourg still led with a combined volume of 177,000 tons, France dropped to 23,000, West Germany provided 47,000, and Japan sent 80,000. In 1986, with a total of 448,000 tons imported, Taiwan provided 117,000 tons; Korea, 52,000; Turkey, 46,000; Singapore, 40,000; Brazil, 30,000; Norway, 28,000; Dominican Republic, 25,000; Trinidad-Tobago, 24,000; and Venezuela, 14,000. Belgium and Luxembourg virtually disappeared from the ranks, as did France and West Germany.

In the first four months of 1987 there were significant changes, with the imports from Taiwan greatly reduced and those from Turkey and Malaysia increased.

This increase in imports between 1984 and 1986, particularly in view of the very low tonnage imported in 1981 and 1982, has been disturbing to the minimills. Further, prices in some areas have been significantly below domestic prices. It is interesting to note that of the 448,000 tons imported in 1986, over 300,000 tons were concentrated in four locations. San Juan, Puerto Rico, received a lion's share with 161,000 tons; Miami, Florida, was second with 78,000 tons; Los Angeles was third with 56,000; and Houston, Texas, was fourth with 27,000 tons. These locations accounted for 232,000 tons of the total. Puerto Rico currently does not have a minimill operating, although there is one located there that has been inactive for some time. It has a capacity to produce about 120 thousand tons.

The prices of imports very often reflect what is necessary to get the business. Frequently, $10 to $15 a ton below the domestic price will suffice. At times, however, the spread has been somewhat larger. The Steel Bar Mills Association, which is made up of minimills, has taken action on this subject in order to remedy the situation.

In the recent past, going back to 1975, minimills have had very little to worry about from imports. However, since 1984, they have felt the pressure in some locations. This is particularly true of those plants in the southeastern section of the United States that can supply rebar to Puerto Rico.

Ease of Entry

Because of the vast difference in the amount of capital investment involved in a minimill contrasted with the integrated steel plant, the ease of entry varies widely.

Entry for Minimills

In the 1960s, when minimills experienced their early growth, the investment in a minimill plant, with an electric furnace, a continuous caster or a breakdown mill for ingots, as well as a bar mill, ranged from under $5 million to $12 million. For example, in 1964 the Tennessee Forging Company at Harriman, Tennessee, built its plant with an annual capacity of 60,000 tons of finished steel products for $4.2 million. Of this amount, $2.8 million was spent on land, buildings, and equipment, including a 20-ton electric furnace with a capacity to melt 100,000 net tons of steel a year, as well as a continuous-casting machine and a relatively small bar mill.

Several other mills were built in the 1960s at a cost of $10–11 million. The possibility of obtaining some second-hand equipment, particularly electric furnaces or bar mills, made it much easier to enter the steel business with a minimum investment. This brought a number of minimills into existence during the 1960s and 1970s.

In the mid- to late 1970s, as inflation increased, the cost of constructing a minimill rose accordingly. Further, the new mills built in the late 1970s were larger, averaging 300,000 tons of raw-steel capacity. Consequently, larger furnaces and rolling mills were necessary and the cost of entry increased to $40–$45 million.

In 1981, Florida Steel built a mill at Jackson in west Tennessee for $60 million. The most expensive minimill built in the United States was Bayou Steel, located near New Orleans. This facility was built by Voest-Alpine, an Austrian steel and engineering company. Voest built an elaborate facility to serve as a showcase in the hope that the company could sell steel minimill equipment in the United

States. The mill's capacity is 650,000 tons of raw steel, of which 200,000 tons were to be sold in the form of billets, since the bar mill did not have the capacity to roll all of the steel that was melted. The investment was in excess of $200 million. Only a company with the financial resources of Voest-Alpine could carry out such a project. Recently, the mill was sold to RSR Steel Company for $76 million. A review of the minimills in operation today indicates that many of them were originally installed for less than $20 million.

Entry for Integrated Plants

In contrast with the ease of entry into the minimill sector of the steel industry, it is much more expensive to break into the industry with an integrated plant. The investment is huge, particularly considering the elements of the integrated mill. United States Steel, in the 1970s, planned to build an integrated mill on Lake Erie. The cost of the 3-million–ton plant was to be $3.6 billion. This is not surprising when one considers that the plant would have to include coke ovens, possibly a sinter plant, a large blast furnace, a basic-oxygen steel shop, a continuous caster, and a rolling mill for either sheets or plates. Even a smaller integrated mill with 1.0 to 1.5 million tons capacity would be very expensive. The cost would be well in excess of $1 billion.

No integrated plant has been built on a greenfield site in the United States by a newcomer to the steel industry in thirty-five years, and it appears that none will be built before the turn of the century. It is possible that entry can be made into the steel business by the purchase of an existing integrated plant, particularly if it has been shut down. This took place in 1986, when the Brenlin Group of Akron, Ohio, purchased the Gadsden steel plant, which was cut adrift from Republic Steel as a condition of the merger with Jones & Laughlin to form LTV Steel.

Entry into Light, Flat-Rolled Products

The investment made by the integrated mills in equipment for the production of light, flat-rolled steel products represents a very large sum. During the 1960s and early 1970s, when the second generation of hot-strip mills were installed, the investment varied from $125 million to $185 million. These mills were referred to as a second generation, since they were far superior to the mills installed before that time. Since 1974, no new hot-strip mills have been constructed. Although the United States Steel division of USX's plan to build a plant at Conneaut on Lake Erie had included a hot-strip mill, the plan was abandoned when cost estimates of the various components showed that the hot-strip mill would cost over $400 million. The latest hot-strip mill installed new is in the Qwangwang plant of Pohang Steel of South Korea, built in 1986 at a cost of $318 million. The investment for the hot-strip mills now in place throughout the United States was considerably less than their replacement cost.

Currently, there are two projects underway which, if brought to successful conclusions, will result in the installation of two new hot-strip mills. One of these is under consideration by Heidtman Company, a steel service center located in Toledo, Ohio. This would be a 1.2-million-ton capacity, 56-inch-wide hot-strip mill. It will be supplied by purchased slabs and require a capital investment of $150 million. The second is a complete plant to be built by Nucor for the production of flat-rolled products. The entire plant, which will include electric furnaces, a continuous caster, a four-stand, 54-inch–wide hot-strip mill, and facilities for cold reduction, will cost $225 million.

In addition to Nucor, four minimills have expressed an interest in entering the light, flat-rolled products field. They are doing so under the assumption that the market for sheets is a large one and entry into it will give them a field for expansion. The desire to expand in this direction is due to the fact that there is an overcapacity to turn out the typical minimill products, such as rebar and small structural shapes. Thus, there is no possibility for expansion in this area. The question arises, however, as to whether or not more capacity is needed to satisfy the demand for light, flat-rolled products. An examination of supply of and demand for light, flat-rolled products since 1977, as well as some indication of the future trends, can shed light on this subject.

In relation to the demand for light, flat-rolled products, an analysis of the past 10 years indicates that there has been a definite decline. Table 4–2 gives domestic shipments of flat-rolled products that are processed through continuous hot-strip mills. In 1977, a total of 48.1 million tons were processed through hot-strip mills. This advanced to 50 million tons in 1978, and almost maintained that total in 1979. In 1980, there was a sharp drop to 39.3 million tons. Since 1981, the level has been approximately 40 million tons with the exception of 1982, when it fell to 32 million tons. Thus, there has been a drop of approximately 10 million tons since the late 1970s.

There are a number of indications that this is a permanent loss. One of these is the decline in demand by the automobile industry for sheet products due to downsizing of cars. In 1976, when 11,497,000 vehicles were produced, the automotive industry purchased 18.2 million tons of light, flat-rolled products. In 1985, when 11,650,000 vehicles were produced, the automotive industry purchased slightly less than 11 million tons of light, flat-rolled products. This sharp reduction in steel purchases, despite the equal production of motor vehicles, is due to the downsizing of cars, the substitution of other materials for steel, and the procurement of parts from outside the country.

The automobile industry is using a light-gauge material that has reduced the sheet tonnage required by 3 percent. It is expected in the years ahead that this industry, which has concentrated in the past few years on the elimination of corrosion, will concentrate more on reducing the weight of the automobile through lighter-gauge, high-strength materials. It has been estimated that by 1992 the sheet tonnage could be reduced by 4 percent to 5 percent. Further, the automakers will also concentrate on reducing the scrap generated, thus improving the yield from steel to finished product and reducing the sheet tonnage required.

In terms of reducing tonnage, a company which is a substantial consumer of cold-rolling sheets reported that by redesigning its equipment and its products with the installation of new automated, computer-controlled facilities, it has reduced the gauge of steel used so that tonnage required is actually some 5 percent less than it was previously. This company has also reduced the amount of scrap generated by about 30 percent. A number of other firms are adopting the same practices, all of which will reduce the tonnage of light, flat-rolled products required. How significant these developments will be in terms of shipments is not easy to determine. However, they will tend to exercise a slight downward bias.

Table 4–3
Hot-Strip Mills over 36 Inches Wide in the United States

Company and Location	*Width*	*Year Built or Rebuilt*	*Capacity (millions of tons)*
Armco:			
Ashland, Ky.	80″	1965R	1.7
Butler, Pa.	58″	1957	.6
Middletown, Ohio	86″	1968	4.3
Bethlehem Steel:			
Burns Harbor, Ind.	80″	1966	3.96
Sparrows Point, Md.	68″	1965R	2.5
Cyclops:			
Mansfield, Ohio	56″	1960R	.720
Ford:			
Dearborn, Mich.	68″	1974	3.0
National:			
Granite City, Ill.	80″	1967	2.7
Ecorse, Mich.	80″	1961	3.4
Inland:			
East Chicago, Ill.	79″	1958R	.6
	80″	1965	4.3
LTV Steel:			
Cleveland, Ohio (J&L)	80″	1964	2.4
Cleveland, Ohio (Republic)	84″	1970	3.8
Warren, Ohio	56″	1961	1.9
Indiana Harbor, Ind.	84″	1968	4.2
McLouth:			
Trenton, Mich.	60″	1967R	1.4
California Steel:			
Fontana, Calif.	86″	1958R	1.5
Sharon:			
Farrell, Pa.	60″	1966R	1.7
Gulf States Steel:			
Gadsden, Ala.	54″	1965R	1.0
United States Steel:			
Irvin Works, Pa.	80″	1963R	2.5
Fairfield, Ala.	68″	1965R	1.9
Fairless Hills, Pa.	80″	1964R	2.6
Gary, Ind.	84″	1967	5.4
Weirton:			
Weirton, W. Va.	54″	1955R	3.0
Wheeling-Pittsburgh:			
Steubenville, Ohio	80″	1965	3.0

Note: R = rebuilt.

higher yield from the slab to the finished coil and also the ability to produce a sheet with a minimum of crown, so that it will have a consistent gauge from side to side. All of the companies operating the second-generation hot-strip mill are examining the possibility of installing this technology. It consists of roll bending, side shifting, and hydraulic screw down, all intended to improve the quality of the sheet. It has been estimated that this technology can be installed for from $2 million to $3 million per stand—$2 million for the newer mills and up to $3 million for the older mills. Many of the second-generation mills have seven finishing stands; thus, the investment is not by any means prohibitive. Most probably these mills will install the technology in the very near future.

The Nucor mill will make use of new technology. It will cast slabs 2 inches thick and reduce them on a hot-strip mill with 4 stands. Space has been left for the installation of a fifth stand, if it is deemed necessary. Production will be 800,000 tons from 2 electric furnaces, 400,000 tons of which will be hot-rolled sheets and 400,000 tons cold-reduced sheets on a single-stand reversing mill. The company will use 250–300 thousand tons of cold-rolled sheets for its own operation; thus, 100–150 thousand tons will be available for sale. Since Nucor uses a small tonnage of hot-rolled sheets, most of the 400,000 tons will be available for sale. The new plant will employ technology that is new and with which there is very little experience. Consequently, there could be some problems, although these are not anticipated. However, it remains to be seen how effective the technology will be and what impact it will have on the cost of production. Since sheets are to be made, the feed for the electric furnace will have to be the highest quality scrap. With the new mill at Nucor, hot-strip capacity will be increased by 800,000 tons and will rise to 65.9 million tons, which will be more than 50 percent above demand.

It is highly unlikely that the market for flat-rolled products will increase to any significant extent in the next decade. The level of consumption in the automotive industry and the container market will most probably remain as it has been since 1984. As a consequence, it is necessary for anyone entering the light, flat-rolled segment of the steel industry to carve out a share of a static market. This means taking a market share from existing producers.

A chief executive of one of the minimills was recently quoted listing favorable factors for the growth of minimills as "new product and process technology that will allow growth into integrated steel markets."[4] The growth referred to here is the growth of the minimills by moving into the integrated companies' product areas. It must be recognized that the integrated companies have the equipment, much of which has been written off. Further, it should be kept in mind that light, flat-rolled products are the heart of the integrated company's business.

In the past, the integrated mills were content to abandon concrete reinforcing bar as well as small structural sections to the minimills. However, the same approach in terms of flat-rolled products is most unlikely, since they represent a major portion of the integrated mills' output. An attempt of a newcomer to penetrate this market will meet with considerable resistance by the integrated companies, most probably in the form of price competition. This could be disastrous. However, a survey of all minimills indicates that very few are interested in entering the light, flat-rolled segment of the steel industry due to the fact that it would represent a costly investment beyond the capability of most minimills.

Notes

1. *American Metal Market,* May 14, 1987, p. 11.
2. American Metal Market Co., *Metal Statistics 1986,* p. 176.
3. T.J. Ess, *The Hot Strip Mill Generation II* (Pittsburgh: Association of Iron and Steel Engineers, 1970), p. 1.
4. *American Metal Market,* April 28, 1987, p. 2.

5
Current Conditions and Prospects

Capacity

The entire steel industry, including integrated plants are well as electric-furnace operations, suffered and still stuffers from overcapacity, due principally to the shrinking market and the increasing penetration of imports. Much has been done, however, to remedy this situation. At present, the official estimate of the American Iron and Steel Institute places capacity to produce raw steel at 112 million tons per annum. This is a drastic reduction from the high point of 160 million tons prevailing during the mid-1970s. The reduction in capacity has allowed the industry to operate at a much higher rate of its potential. In spring 1987, operating rates were as high as 83 percent of capacity, a figure that had not been attained since the late 1970s. Despite the improvement, there is need for a further reduction in capacity of at least 15 million tons. However, it will probably be no more than 9–12 million tons. This will put the capacity ultimately at 100–103 million tons, which should be quite adequate for steel production of 85–90 million tons, a figure that will probably not be surpassed for some length of time during the next decade.

Raw-steel output recently went from a low of 74.6 million tons in 1982 to a high of 92.5 million tons in 1984, with a subsequent drop to 80.5 million tons in 1986. Part of the reason for the decline in 1986 was the rise in continuous-casting capability during the past two years, increasing the yield from raw steel to finished product; requiring less raw steel. There will be some additional continuous-casting capacity installed in 1987–88, which will bring the total figure to more than 70 million tons. At this level, it is doubtful that much more continuous-casting capacity will be needed.

The reduction in capacity has contributed, to a significant extent, to the recent stability in prices, although the improved conditions of the past six months have also been caused, in great part, by the six-month strike sustained by the United States Steel division of USX. This strike removed 16–17 percent of the country's capacity from the market and permitted the remaining companies to increase their output. With this reduced capacity, prices, particularly for light, flat-rolled products, became firm during the latter part of 1986 and were increased in the first quarter of 1987. When United States Steel came back into the market at the termination of the strike on February 1, 1987, many customers expected that United States Steel would cut the going price in order to recoup its share of the market. This did not happen. A number of customers waited for a month in the hope that there would be price reductions and, consequently, reduced their inventories appreciably. When the price reductions did not come, they had to place orders at the higher price in order to replenish their inventory. This was particularly true of light, flat-rolled steel. Thus, the increase in demand, coupled with a firm price, produced a profitable situation for most of the integrated companies in the first quarter of 1987. This continued through the second quarter and hopefully will result in a profitable year for many of the integrated steel producers.

As indicated, there will be future capacity reductions which will probably be achieved through the partial closure of some mills, although it is conceivable that one or two entire plants may be closed.

Integrated Mills

In this book's treatment of integrated mills, the closures in the past several years have been detailed. In spite of these closures, the finishing capacity for a number of products remains more than adequate. The principal products of the integrated producers consist of varying types of sheet and strip, including hot- and cold-rolled sheets as well as coated materials, such as tinplate and galvanized steel.

In 1986, some 40 million tons of these products were shipped, all of which were processed through hot-strip mills. The hot-strip mill capacity currently stands at 65.1 million tons and is much more than adequate to handle the current demand and even an increase of 20 percent, which is rather unlikely.

In terms of other facilities, plate mill capacity is far in excess of demand. In 1986, some 3.6 million tons of plates were shipped by the domestic producers, which by no means taxed the plate mill capacity of the country, which is 7–8 million tons. Rail mill capacity stands at 1.1 million tons, if the 3 mills, including the installation at Monessen, are counted. This was far more than needed to supply the 461,000 tons of rails shipped in 1986—an all-time post–World World II low.

In the next decade, there is no need for additional finishing facilities in terms of providing more tonnage. However, the strip mills will have to be upgraded so that they can provide a better quality sheet for the consumers, since many of them are demanding product with specificiations in excess of the capabilities of some strip mills. Ironically, the demand for higher-quality sheets in virtually all categories will lead to a reduction in the tonnage, since many consumers are using lighter sheets.

The time-honored uses of flat-rolled products in a number of instances will witness a reduction in tonnage. This will have an effect on the production of light, flat-rolled products and force a contraction in the market unless new applications are found. A great deal of effort has been and is being expended in this direction.

The trend to less tonnage will make entrance into the light, flat-rolled product market less attractive for any company installing new capacity, since they will have to compete vigorously to attain a share of the market. This would be particularly true if the total amount of new capacity exceeded 2 million tons.

Minimills

Nucor expects to produce 800,000 tons of sheets, half of which will be cold-rolled. Thus, it will have some 400,000 tons of hot-rolled sheets to put on the market and a relatively small amount of cold-rolled, since it will use most of the 400,000 tons that it will produce. However, this means that the companies from which its cold-rolled purchases were made will have that extra tonnage to sell. Thus, additional capacity, brought into a static market, could result in price competition. If the capacity is large enough, the competition could bring prices down to a point where they might be below costs.

In the decade ahead, the need for seamless pipe for oilwell operations can only grow if the price of oil reaches at least $25 a barrel or if a shortage of oil develops over a protracted period due to disturbances in the international market. This could increase drilling appreciably and would result in a higher demand for oil-country tubular goods. In 1986, demand reached an all-time post–World War II low of 483,000. Should it recover to 2 million tons or more, there would still be adequate capacity to handle it, particularly in view of the fact that half of this tonnage is in the form of welded pipe and the other half seamless pipe.

There is more than enough capacity to handle welded pipe. Lone Star Steel alone has some 800,000 tons of welded capacity, and there are a number of smaller operations, many of them nonintegrated, that supply this product. In terms of seamless pipe, United States Steel has a capacity to produce 600,000 tons at its new mill in Fairfield, Alabama, as well as several hundred thousand tons more at its Lorain, Ohio, facility. North Star, in Youngstown, has added some 300,000 tons, while CF&I Steel can contribute significant tonnage. Thus, the 1 million tons of seamless oil-country tubular goods can be more than adequately cared for by existing facilities.

Imports have absorbed 55 percent of the market and will maintain that rate so that if, by 1992, total demand for oil-country tubular goods exceeds 4 million tons, the domestic portion will be about 2 million tons.

The possibility of additional small electric-furnace plants entering the seamless business is very remote. One attempt was made in 1982, when York-Hanover's effort to raise the necessary capital to build an electric-furnace plant to feed a modern seamless mill with 250,000-ton capacity was not successful.

The market for large structural sections during the next decade will depend on the rate of high-rise construction as well as major construction projects, such as bridges. It has been projected to be relatively static, and there will be spirited competition for the tonnage. Currently, there are 5 companies that can produce wide-flange beams. These include United States Steel and Bethlehem Steel, whose mills can roll sections up to 36 inches wide. Inland has a capability to roll up to 24 inches in width, while Northwestern Steel and Wire can roll up to 18 inches. The newest entry is Chaparral with a capability to roll to 14 inches in width. Nucor, with its joint venture partner, Yamato Steel of Japan, is building a structural mill in Arkansas on the Mississippi River, with a capacity for 650,000 tons, of which 100,000 tons will be railroad tieplates and the remainder wide-flange beams up to 24 inches wide.

In addition to the domestic producers, Nippon Steel in Japan, British Steel in the United Kingdom, and Arbed Steel in Luxembourg have shipped considerable tonnage of wide-flange beams to the United States. In 1986, among the three, total shipments were over 400,000 tons.

Chaparral may well build another plant for the production of structural shapes. This will add more tonnage to supply a market that has been relatively stable and is projected to remain so during much of the next decade, particularly considering that some cities have overbuilt their high-rise office buildings. The mill that Nucor has under construction can, by no means, be classified as a minimill because of its tonnage and type of product. Nucor will have a larger capacity for wide-flange beams than some of its competitors. Between Nucor and Chaparral (two companies that are classified as minimills), there will be more than a million tons of capacity, thus assuring them a dominant position in this field.

Raw Materials

Integrated Mills

In the 1990s, some of the integrated steel companies will cease to own huge deposits of raw materials and will rely to a greater extent on purchasing them. With the drop in steel production in recent years and the projection of lower output in comparison with the 1970s, integrated companies find that they are oversupplied with raw materials. There is also much greater opportunity to purchase raw materials, since large deposits of iron ore have been developed in Canada, Brazil, Venezuela, Australia, and West Africa. The time-honored practice of full integration back through raw materials is no longer considered necessary. In fact, what was once essential, particularly in boom times, has now proved to be a liability, particularly since richer ores are available at low prices.

In regard to metallurgical coal, a number of integrated companies have sold off large deposits obtained in the early years of the twentieth century. They find it advantageous to buy coal that is readily available in the United States. For some companies, this practice extends through coke, principally because of the large investment required to rebuild batteries of coke ovens.

Recently, a decision has been made by National Steel not to rebuild a coke-oven battery at its Great Lakes plant because of the $60 million investment required. Since coke is available from a number of sources at reasonable prices, the decision was made to use the $60 million for other equipment. Further, the amount of coke required by the industry has declined significantly with the production of less than 50 million tons of pig iron annually since 1982, while, with the improvement in the coke rate, there has been a sharp decline in the demand for coke, and there is little prospect that this decline will be reversed significantly in the decade ahead.

In the 1990s, the integrated companies will purchase more raw materials than they have in the past.

Minimills

The minimills will continue to use scrap as a raw material. In some instances, where flat-rolled products will be produced, there will be a need to upgrade the electric-furnace charge with some direct-reduced iron ore or high-grade selected scrap.

Minimills, because of the extremely high yield of end product from raw steel, have very limited amounts of mill or revert scrap. Further, because of the products they make, there is limited demand for higher-priced prompt industrial scrap. The major portion of their scrap comes from obsolete metal; witness the fact that several minimills draw 100 percent of their supply from scrap dealers with which they have an association. Some mills also operate scrap shredders.

Since most minimills will continue to produce the traditional products, such as rebars and small structural sections, they will continue to use the same grades of scrap that have fed their furnaces for a number of years. In moving into the more sophisticated products, such as seamless pipe and sheets, there will be a need, as indicated, to upgrade raw materials. However, the number of companies that would be involved in these operations is extremely small. Therefore, for the minimills in general, scrap will not present a problem. The need to upgrade the charge for the electric furnaces on the part of those companies venturing into flat-rolled products could require the location of these plants in areas where prompt industrial scrap is readily available.

Investments

The integrated companies during the past few years have concentrated the major portion of their capital expenditures on continuous-casting units and finishing facilities. The continuous-casting program, with one or two exceptions, will be completed by the end of 1987, when a total of over 70 million tons of continuous-casting capacity will have been installed. This is sufficient to cast 80 percent of the steel that will be produced in the next decade.

Attention will be given in the next few years to such items as ladle furnaces (which improve the quality of the steel) and finishing equipment (such as hot-strip mills and cold-strip mills). The hot-strip mills will be upgraded with such items as roll bending and side shifting as well as hydraulic screwdowns, which will permit the production of a much better coil, particularly in regard to consistency of gauge and width.

A considerable investment is being made at present in two new cold-strip mills that are under construction. One at Inland Steel is a joint venture between Inland and Nippon Steel of Japan; the other is at Pittsburgh, California, where a joint venture has been established between United States Steel and Pohang Iron and Steel Company of South Korea. These mills are superior to existing cold-reduction mills, particularly insofar as each stand will have six rolls in place of the conventional four, and there will be improved annealing and pickling facilities. All of this will result in a much better quality cold-reduced sheet. The production of this superior product will require other integrated steel companies to install similar equipment in order to compete. The investment in these cold-strip mills is in excess of $400 million in each case. The Inland faciliy has been financed, in great part, by Japanese trading companies, while the United States Steel and Pohang joint venture is financed in the conventional manner. It is conceivable that by the mid-1990s, there will be three or four more of these units installed by other steel companies.

Minimills will continue to invest in their existing facilities by either improving them or replacing them. This will result in some expansion as well as an improvement in the quality of the product and a reduction in production costs. Those mills that are operating cross-country bar mills or combination cross-country and in-line mills will more than likely shift to straight in-line mills. Reheating furnaces and cooling tables will be replaced and improved, and the basic electric furnaces will be upgraded in terms of increased power and, most probably, eccentric bottom tapping. In addition, there will be some investment in downstream facilities, which will permit the mill to further fabricate its product.

Reductions in Capacity

Most attempts to reduce capacity, if they involve the closure of an entire plant or a major facility in the plant, have met with difficulties. First, if the pensions of the employees that will be separated are not fully funded, it is necessary for the company to write off the pension cost of the facility involved. Most integrated companies do not have fully funded pensions, with the exception of United States Steel and Inland Steel. Thus, closing a plant becomes a very expensive proposition. The amount that must be written off can run into many millions of dollars; in some instances, the companies feel it is less expensive to run the plant at a moderate loss than to take a huge pension write-off. As a consequence, a number of facilities that should have been abandoned are still in operation, and this excess capacity exercises a downward bias on prices, since it maintains supply in excess of demand. There has been much concern about this problem, but no satisfactory solution has as yet been reached.

The second difficulty in closing facilities, particularly if an entire plant is involved, is the attempt by the employees and that of the community to revive the plant in order to preserve jobs. The community's interest is of great importance, since the presence of a functioning steel mill provides revenues from taxes and accounts for other jobs for suppliers of goods and services to the plant.

Thus, there is a dilemma. On the one hand, the capacity must be reduced when one considers the overall good of the steel company and the steel industry. If the plant were considered necessary, the company obviously would not have to close it. It is marked for closure usually because it represents unneeded capacity and very often has obsolete, high-cost facilities. Thus, considering the needs and requirements of the industry, the plant should be closed. On the other hand, the plant represents the jobs and the livelihoods of the employees involved, and they are understandably interested in preserving their jobs. Similarly, the community is understandably interested in preserving the business the plant provides. As a consequence, there will almost always be an attempt to revive a steel mill that has been shut down. This applies not only to integrated mills, but also to minimills.

The employees in a number of instances have tried to purchase a plant and operate it as an Employee Stock Ownership Plan (ESOP). If this fails, an investor is very often brought in who may purchase the facility with a minimum capital investment. In addition, a labor contract is often drawn up with the employees providing much lower wages than were received before the plant was closed. As a consequence, the restructured mill, with its lower capital investment and significantly lower labor costs, has a competitive advantage over other steel mills.

A case in point is the Gulf States Steel Corporation in Gadsden, Alabama. This was formerly a plant of the Republic Steel Corporation and was to be disposed of in order to allow the Republic and Jones & Laughlin merger. The employees attempted to establish an ESOP without success. Failing this, the Brenlin Group of Akron, Ohio, acquired the plant and arranged for a labor contract which was some $6 per hour less than that in force at a steel mill some 60 miles away. Another example is the McLouth Steel Corporation, whose name has recently been changed to McLouth Steel Products Corporation. Faced with bankruptcy, the company, in chapter 11, negotiated a settlement with the union by which the employees agreed to take a $5 per hour reduction in pay. When this did not save the company and it was on the verge of liquidation, Cyrus

Tang purchased it and it continues to operate. However, recently Cyrus Tang offered to sell his interest in the company to the employees.

Not all efforts at revival have been successful. One that did not succeed was the attempt to restart the Republic Steel Corporation plant in Buffalo, New York. This was written off in 1984 just prior to the merger with Jones & Laughlin. However, the plant had been inoperative for a year or more prior to the write-off. As a consequence, its market was lost. The lieutenant governor's office in New York State initiated a plan calling for a study to determine whether or not the plant could be reopened. The study was subsequently carried out, and the results were negative due to the fact that the market had been lost and many of the facilities were obsolete. Thus, 1 million tons were taken out of the capacity of the steel industry.

In regard to minimills, a number of these have been closed and reopened. This is a much easier task, since the investment involved in purchasing a closed minimill plant is usually very much less than that required for a closed integrated mill. The most recent example is Tennessee Forging, which was closed, has been purchased, and is scheduled to begin operation in late 1987. Another example is Kentucky Electric, a plant closed as a result of a labor dispute, but subsequently purchased by Newport Steel Corporation and put back in operation.

To avoid these difficulties involved in plant closures, a number of companies have reduced capacity by shutting down certain parts of their plants. For example, at the Great Lakes plant of National Steel, one of two basic-oxygen steelmaking shops was closed, thereby reducing capacity by some 40 percent. Since the remainder of the plant still operates, there is no attempt to reopen the closed facility.

Conclusion

Both the minimills and the integrated mills are essential to the steel industry as it exists currently and will continue to exist through the end of the century. The integrated mills are large and expensive to construct. This is unavoidable if iron ore is to be reduced to pig iron and further refined into steel. Currently, for a large production of 50 million tons of iron, blast furnaces are needed. They are expensive and must also be supported by coke ovens to produce the coke required for the reduction of ore to pig iron, but the job cannot be done on the scale needed without them. There have been some processes, such as the KR process, that bypass the blast furnace; however, they can only produce small amounts of pig iron.

Further, the demand for steel, which, in 1986, required 80.5 million tons of raw steel, can only be satisfied in part with large units—plants capable of producing several million tons. Likewise, the production of 40 million tons of light, flat-rolled products requires huge facilities. It would be impractical and uneconomical to think of a series of small mills, scattered throughout the country, producing 700,000 to 800,000 tons of light, flat-rolled products, rather than relatively few plants that can produce several million tons. Thus, the steel industry, if it is to produce economically what is needed, must operate large integrated plants. Further, the steel produced from raw materials, such as iron ore and coal, is relatively free of contaminants, and is needed to produce certain products such as deep-drawing sheets.

The minimills, during the past two decades, have established themselves as a necessary part of the steel industry, becoming almost the only producers of a number of products. These include rebar and small structural sections. The type of steel made by the minimills, using scrap as a raw material, is more than adequate for the type of products they make. Minimills have 17 percent of the industry's capacity, and this may increase slightly in the years ahead, although the capacity to produce minimill-type products is more than adequate for the foreseeable demand.

There has been considerable discussion about the possibility of the minimill or a small plant based on electric furnaces penetrating the market usually considered the domain of the integrated producers. The Nucor mill, currently under construction to produce medium-wide sheets, will be watched carefully, since it will employ new technology. A proliferation of this type of mill would be difficult. To begin with, the mill is based on scrap, and although one unit with 800,000 tons of capacity could collect enough selected scrap or supplement the scrap with either pig iron or directly reduced iron, it is almost inconceivable to think of a large number of these mills being able to gather enough quality scrap or direct-reduced iron to produce high-grade steel sheets.

Further, the market for light, flat-rolled products is by no means large enough to use current capacity in place. Remember, there is at least a 50 percent overcapacity in terms of strip mills producing light, flat-rolled products. A large infusion of new capacity would most probably result in severe and destructive price competition and an uneconomical use of capital.

To a certain extent, this prospect seems to be academic, since the vast majority of the minimills interviewed in connection with this book have no intention of entering the light, flat-rolled products sector of the steel business. Many of the minimills that have operated profitably are quite content to continue with the products with which they have found a niche.

The integrated mills, slimmed down and devoid of most of their high-cost and obsolete facilities, have improved productivity and reduced costs. Given an increase in demand of relatively minor proportions, they can return to profitability.

The comparison between the minimills and the integrated plants indicates the need for both these components of the steel industry to satisfy the demands of the economy, with each group producing products for which its facilities are best adapted. The integrated plants have moved out of the area covered by the minimills, and although a few minimills will move into the product areas of the integrated plants, there will not be a surge in this direction.

Interviews with management of both integrated and minimills indicate very little change in the product areas, although there will be changes in technology and corporate organization. Both sectors of the steel industry will continue to improve their plant and equipment, and some will restructure their corporate organization. The integrated plants have no intention of moving back into the rebar and small structural product lines, and the attempt by two minimills to move into sheets and seamless pipe will be limited for a number of years to those companies.

Statements have been made that the technology is available for casting thin slabs, which will tempt a number of minimills to adopt it. This has not been borne out in discussions with the minimills.

Nucor will have its plant by 1988 or early 1989, but it will be at least 1992 or 1993 before anyone else in the minimill field attempts to follow Nucor's lead. The same available technology can be used by the integrated plants—and possibly will be—but this will not be attempted before the early 1990s. Thus, with few exceptions, the minimills will keep to their niche, as will the integrated plants, for quite a few years to come.

Bibliography

American Iron and Steel Institute. *Annual Statistical Reports*. Washington, D.C.: AISI, various years.

American Metal Market Co. *Metal Statistics: The Purchasing Guide of the Metal Industries*. New York: AMM, various years.

"American Steel: Resurrection," *The Economist* (April 2, 1983): 75–76.

Barnett, Donald F., and Crandall, Robert, W. *Up From the Ashes: The Rise of the Steel Minimill in the United States*. Washington, D.C.: The Brookings Institution, 1986.

Center for Metals Production. "Electric Arc Furnace Steelmaking . . . The Energy Efficient Way to Melt Steel," *CMP Tech Commentary*, vol. 1, no. 3, 1985.

Cordero, Raymond, and Serjeantson, Richard, eds. *Iron and Steel Works of the World*, 8th ed. London: Metal Bulletin Books, Ltd., 1983.

Editors of 33 Metal Producing. "The American Mini-Mill at Midstream," *33 Metal Producing* (February, 1987): 27–41.

Ess, T.J. *The Hot Strip Mill Generation II*. Pittsburgh; Association of Iron and Steel Engineers, 1970.

Geiger, Gordon H. "Mini-Mills, Technology, and People," *ASM Metals Congress*. Detroit: September 19, 1984.

Hogan, William T., S.J. *Steel in the United States: Restructuring to Compete*. Lexington, Mass.: Lexington Books, D.C. Heath & Co., 1984.

Huettner, David. *Plant Size, Technological Change, and Investment Requirements*. New York: Praeger Publishers, 1974.

Innace, Joseph J. "The Long, Quiet Twilight of the American Blast Furnace," *33 Metal Producing* (November, 1985): 41–47.

Iron and Steel Society. "Electric Arc Furnace Roundup—U.S.A.," *Iron & Steelmaker* (May, 1987): 18–39.

Isenberg-O'Loughlin, Jo, and Innace, Joseph T. "Full Steam Ahead on the Nucor Unlimited," *33 Metal Producing* (January, 1986): 35–49.

Koelble, Frank. "Strategies For Restructuring the U.S. Steel Industry," *33 Metal Producing* (December, 1986): 28–33.

Kotch, J.A. "Neighborhood Steelmaking: A Look at the Mini-plants," *Iron and Steel Engineer Year Book* (1971): 411–428.

Kredietbank. "Mini Steelworks: Pygmies among Giants," *Weekly Bulletin* (March 27, 1981): 1–5.

Labee, Charles J., and Samways, Norman L. "Developments in the Iron and Steel Industry, U.S. and Canada—1986," *Iron and Steel Engineer* (February, 1987): D1-D24.

Marcus, Peter F., and Kirsis, Karlis M. *Economics of the Mini-Mill: World Steel Dynamics Core Report X.* New York: PaineWebber, 1984.

McManus, George J. "Low Costs and High Quality: The One-Two Punch of the Electric Furnace," *Iron Age* (January 3, 1983): MP7-MP17.

McManus, George J. "Now the Slump Threatens to Pinch the Minimill Growth," *Iron Age* (August 2, 1982): MP5-MP15.

"Mixed Prospects for U.S. Minis," *Metal Bulletin Monthly* (March, 1983): 7-11.

Nemeth, Edward L. "Mini-midi mills—U.S., Canada and Mexico," *Iron and Steel Engineers* (June, 1984): 25-56.

Peters, A.T. *Ferrous Production Metallurgy*. New York: John Wiley & Sons, 1982.

Russell, Clifford S., and Vaughan, William J. *Steel Production: Processes, Products, and Residuals*. Baltimore: The Johns Hopkins University Press, 1976.

Schroeder, D.L. "Computer Systems in Mini Mills," *Iron and Steel Engineer* (November, 1984): 57-61.

Sendizimir, M.G. "Hot Strip Mills for Thin Slab Continuous Casting Systems," *Iron and Steel Engineer* (October, 1986): 36-43.

Sheets, Kenneth R. "U.S. Minimills Getting Caught in Steel Trap," *U.S. News & World Report* (December 9, 1985): 51.

Stone, J.K., and Michaelis, E.M. "LD Capacity Stagnant: Process Evolution Continues," *Iron and Steel Engineer* (September, 1984): 27-34.

Strohmeyer, John. *Crisis in Bethlehem: Big Steel's Struggle to Survive*. Bethesda, Maryland: Adler & Adler, 1986.

U.S. Congress, Office of Technology Assessment. *Technology and Steel Industry Competitiveness*. Washington, D.C.: U.S. Government Printing Office, 1980.

Wheeler, Frank M., Leon, Henry W., and Yates, J. Roger. "The Mini Mill Faces New Technologies," *Iron and Steel Engineer*, (January, 1986): 53-55.

Wilshire, B., Homer, D., and Cooke, N.L. *Technological and Economic Trends in the Steel Industries*. Swansea, United Kingdom: Pineridge Press Ltd., 1983.

Index

About the Author

Rev. William T. Hogan, S.J. received his Ph.D. in economics from Fordham University in 1948. He has conducted economic studies of the steel industry and other basic, heavy industries for the past thirty-five years. During this time he has authored a number of books, including *Productivity in the Blast-Furnace and Open-Hearth Segments of the Steel Industry*, the first detailed study of steel productivity; *The Development of American Heavy Industry in the Twentieth Century*; and *Depreciation Policies and Resultant Problems* (1967). His five-volume work, *Economic History of the Iron and Steel Industry in the United States* (Lexington Books, 1971) covers industry developments from 1860 to 1971. The companion studies to this are: *The 1970s: Critical Years for Steel* (1972), *World Steel in the 1980s: A Case of Survival* (1983) and *Steel in the United States: Restructuring to Compete* (1984).

In 1950, Father Hogan inaugurated Fordham University's Industrial Economics Research Institute, which has produced numerous studies dealing with economic problems of an industrial nature. He has appeared before legislative committees of both the U.S. Senate and the House of Representatives and has testified several times before the House Ways and Means Committee on legislation affecting depreciation charges and capital investment. He has also appeared before the Senate Finance Committee to testify on tax incentives for capital spending. He was a member of the Presidential Task Force on Business Taxation and a consultant to the Council of Economic Advisers to the President and the U.S. Department of Commerce.

During the past twenty years, Father Hogan has visited most of the steel-producing facilities in the world and has delivered papers at steel conferences in France, United Kingdom, Switzerland, Sweden, Czechoslovakia, Russia, Venezuela, Brazil, South Africa, India, the Philippines, South Korea, and Japan. He is the author of numerous articles on various aspects of steel industry economics.

In 1985, Father Hogan was awarded the Gary Memorial Medal, which is the highest honor the steel industry can bestow.